SpringerBriefs in Electrical and Computer Engineering

For further volumes:
http://www.springer.com/series/10059

Xin Wang

Scheduling and Congestion Control for Wireless Internet

 Springer

Xin Wang
Department of Computer and Electrical
  Engineering and Computer Science
Florida Atlantic University
Boca Raton, FL
USA

ISSN 2191-8112   ISSN 2191-8120   (electronic)
ISBN 978-1-4614-8419-6   ISBN 978-1-4614-8420-2   (eBook)
DOI 10.1007/978-1-4614-8420-2
Springer New York Heidelberg Dordrecht London

Library of Congress Control Number: 2013943707

Printed on acid-free paper

Springer is part of Springer Science+Business Media (www.springer.com)

To Na, Nancy, and Nathan

Preface

Most of the practical network protocols were designed based on sound yet ad-hoc heuristics. They have performed reasonably well as the Internet scaled up by six orders of magnitude in size, speed, load, and connectivity over the past three decades. With the Internet continuing to grow and include an increasing number of wireless links, however, it is well acknowledged that the current protocols need to be re-engineered or optimized for emerging wireless applications.

Both engineering heuristics and optimization-theoretic approaches were recently employed to enhance or re-engineer the existing network protocols. The resultant schemes, however, either cannot provide analytical performance guarantees, or are difficult to be deployed with the current Internet infrastructure. To fill the gap, we develop a systematic theory-guided design approach to *readily deployable, scalable, yet optimal* network schemes for *wireless Internet applications*. As core of the Internet protocols, the Transmission Control Protocol (TCP) performs window-based congestion control for competing data flows. From optimization-theoretic perspective, TCP window-control mechanism amounts to implicit updates of source rates as "primal variables" and queueing delays as "Lagrange dual variables". Confined to the design space of Internet, we propose that the key for cross-layer optimization of Internet protocols is to develop non-standard *window-based implicit primal-dual solvers* for underlying optimization problems, and rely on queueing delays to *decompose* these solvers into local algorithms that can be deployed and operated asynchronously at different layers of distributed network nodes.

This book is devoted to establishing the existence of such solvers and creating a systematic framework for their design and analysis, with a focus on joint design of link-scheduling (at link layer) and congestion control (at transport layer) for wireless Internet applications. In Chap. 2, we capitalize on the proposed approach to prove the existence of *readily deployable, scalable, yet optimal* joint congestion control and wireless-link scheduling schemes for TCP flows over the Internet with cellular wireless links. In Chap. 3, we generalize our approach to joint optimization of TCP congestion control and distributed link-scheduling for the Internet traffic over ad-hoc wireless links. The design and analysis approaches in Chaps. 2 and 3 can be simply combined to provide the solutions for the Internet with both cellular and ad-hoc wireless links. In Chap. 4, we outline some possible generalizations and interesting directions for the proposed approach. Chapter 5 summarizes the book.

As an extension, integration, and re-engineering of the current state-of-the-art theory-guided network design methods, the proposed novel approach is an attempt to bridge the network optimization theory and practical Internet designs, by developing non-standard optimization algorithms within the Internet design space toward network protocols/schemes that can be used with current infrastructure. Generalization of the approach is possible to embody a paradigm shift in the theory and practice for cross-layer design and optimization of Internet protocols. As a result, the research can impact the *basic theory* in understanding and optimizing the large-scale network and communication systems, and is expected to benefit directly *applications* to next-generation Internet protocol designs. In an even broader sense, the high-performance network schemes for wireless Internet applications can have an *impact on society*, if one takes into account how the Internet and wireless devices are expected to transcend various aspects of our everyday life.

Contents

1 Introduction 1
   1.1 Current State of Knowledge 1
      1.1.1 Network Utility Maximization Paradigm 1
      1.1.2 Window-Based Congestion Control 2
   1.2 The Gap 3
   1.3 The Approach 4
   1.4 Outline of the Book 5
   References 7

2 Joint Congestion Control and Link Scheduling for Internet
   with Cellular Wireless Links 11
   2.1 Network Model and Problem Formulation 11
   2.2 Algorithm Development and Analysis 13
      2.2.1 Joint Design in Network Fluid Model 13
      2.2.2 Convergence Analysis 15
      2.2.3 Practical Algorithm Development 17
   2.3 Summary 18
   References 22

3 Joint Congestion Control and Link Scheduling for Internet
   with Ad-Hoc Wireless Links 23
   3.1 Network Modeling, Problem Formulation,
       and Existing Solutions 24
      3.1.1 Idealized CSMA Protocol 24
      3.1.2 Problem Formulation 26
      3.1.3 Queue-Length Based Solutions 27
   3.2 Joint Congestion Control and CSMA Scheduling
       Without Message Passing 30
      3.2.1 Joint Design in Network Fluid Model 30
      3.2.2 Convergence Analysis 32
      3.2.3 Development of Practical Schemes 33
   3.3 Summary 34
   References 37

4   Generalizations and Interesting Directions   39
     4.1   Internet with Both Cellular and Ad-Hoc Wireless Links.   39
     4.2   General Utility Functions   41
     4.3   Interesting Directions.   43
            4.3.1   Impact of Feedback Delays and Network Dynamics   43
            4.3.2   Multi-Path Routing   44
     4.4   Summary   47
     References   51

5   Summary.   53

Chapter 1
Introduction

Most of the practical Internet protocols were designed based on sound yet ad-hoc heuristics. They have performed reasonably well as the Internet scaled up by six orders of magnitude in size, speed, load, and connectivity over the past three decades [1]. However, many studies have shown that standard Internet protocols perform poorly in networks with large bandwidth-delay products and/or lossy wireless links [2–4]. To meet the growing demand for ubiquitous data services, we have witnessed a rapid convergence of (wireless) communication, computation, and control systems into a global Internet. As the Internet continues to grow and extends from its wired backbone to include an increasing number of wireless links, it is well acknowledged that current protocols need to be re-engineered or optimized for emerging wireless applications.

1.1 Current State of Knowledge

Network protocols are critical, yet difficult to understand and optimize, since they consist of *local* algorithms that are distributed spatially and vertically, and operated asynchronously to accomplish a certain *global* goal. Despite the difficulty, great progress has been made in the last decade. Both engineering heuristics and optimization-theoretic approaches were developed to enhance or re-design the existing network protocols. Next we briefly review the current state-of-art in the field of network design and optimization.

1.1.1 Network Utility Maximization Paradigm

To understand the large-scale Internet under end-to-end control, a notable duality network fluid model was developed [5–7], where the Transmission Control Protocol (TCP) is interpreted as a distributed algorithm to solve a global optimization

X. Wang, *Scheduling and Congestion Control for Wireless Internet*,
SpringerBriefs in Electrical and Computer Engineering,
DOI: 10.1007/978-1-4614-8420-2_1, © The Author(s) 2014

problem. Different TCP algorithms differ in the objective function of the underlying optimization and the iterative operation to solve it. Based on this idea, a network utility maximization (NUM) paradigm was proposed to re-engineer the existing network protocols and structures via convex/nonlinear optimization tools. The NUM paradigm has become a popular approach to modeling, analyzing, and designing resource allocation schemes across layers for wired or wireless networks [8, 9]. Under the umbrella of NUM, joint congestion and power control schemes were developed in [10, 11], joint congestion and contention control were explored in [12–14], joint congestion control and scheduling were proposed in [15, 16], joint routing and power control were investigated in [17–19], joint congestion control, routing, and scheduling were addressed in [20–22], and joint routing, scheduling, and power control schemes were put forth in [23]; see also tutorials [24–26].

In most NUM schemes, congestion control is simplified as a direct source-rate controller, which can be classified as dual-controller [9, 16, 20, 27, 28] or primal-dual controller [8, 21, 29–31]. In both dual and primal-dual controllers, queue lengths of links play the role of (scaled) Lagrange multipliers. The proposed source rate updates and queue evolutions consist of a classic gradient-type dual or primal-dual iteration to solve the underlying NUM problem. This approach facilitates cross-layer optimization of routing, scheduling, medium access control (MAC), and/or power control schemes. For instance, it was shown that queue backpressure and maximum weight (MaxWeight) matching based dynamic routing and scheduling algorithms can maintain network stability whenever possible [18, 32]. With queue lengths playing the role of Lagrange multipliers, the source-rate control and such a queue backpressure mechanism can be seamlessly "glued together" as the decomposition parts of a global gradient-type iteration to develop joint congestion control, routing and scheduling schemes that maximize the aggregate network utility subject to stability [20, 27, 30]. Joint designs of congestion control, MAC, and/or power control schemes can be also derived following similar decomposition of the global primal-dual or dual-based iterative solutions to the relevant NUM optimizations [10, 12, 13, 24, 33].

Since the development of the NUM schemes follows the classic gradient-type algorithms, the stability, convergence, and optimality of the resultant algorithms can be readily shown by drawing from the standard convex analysis and stochastic approximation tools. This approach is mathematically tractable and appealing. As a result, it has attracted growing interest in network design and optimization research.

1.1.2 Window-Based Congestion Control

In TCP congestion control [34–37], each flow source adjusts its transmission window size (i.e., the maximum amount of outstanding packets that it can send to the network) based on its own observations of network congestion. Locally observable congestion measure (packet loss or delay) is employed to determine window adjustment. Self-clocking is enforced so that the source sends a new packet after receiving an acknowledgement (ACK) packet when its window size is fixed, or it sends out bulk traffic in bursts upon change of window size. In this way, each source always maintains the number of packets in-the-fly equal to its window size, and packet transmission is clocked at the same rate as the throughput that the flow receives.

Engineering heuristics were adopted to enhance the TCP window-control for large-scale Internet with wireless links. Poor performance of the standard TCP (i.e., TCP-Reno) over wireless is partly due to its additive-increase-multiplicative-decrease (AIMD) window adjustment based on the packet loss that is assumed to be caused solely by network congestion [3]. Building on sound heuristics, some new window adjustment algorithms were put forth to increase window sizes more aggressively and decrease less drastically in HSTCP and STCP [38, 39]. A number of schemes were developed to make TCP more resilient by distinguishing between packet loss due to congestion and that due to wireless-link errors in ATCP, TCP-Veno, TCP-Westwood and TCP-Jersey [40–43], or by reducing/controlling the link-layer packet loss rates to a tolerant level [4, 44–47]. Some active queue management and/or receiver-side control schemes were also proposed to enhance TCP; see CLAMP, Freeze-TCP, and others [43, 48–51]. Simulations and experimental measurements were used to evaluate theses schemes.

Recently, it was increasingly recognized that queueing delay provides more accurate information on network congestion than packet loss, and can help maintain stability as network capacity grows [2, 52]. This motivates the TCP-Vegas, TCP-Westwood, TCP-LP, TCP-Veno, TCP-Jersey, BIC-TCP, CUBIC, CTCP, etc., which employ (in part) queueing-delay based solutions for congestion control [41–43, 53–58]. In a more systematic manner, queueing-delay based congestion controllers in Mo-Walrand scheme and FAST-TCP [2, 59] were designed and analyzed for wired Internet under the network fluid model, where queueing delays were used to play the role of Lagrange multipliers for underlying network optimization. Stability/convergence of the proposed schemes were either analytically established using a Lyapunov method [59], or demonstrated by extensive simulations and experiments [2].

1.2 The Gap

Both the existing NUM solutions and TCP enhancements have limitations. In the NUM paradigm, simplifying congestion control as a source-rate controller and using queue lengths as congestion measures facilitate development of classic gradient-type solutions. However, the source-rate controllers in NUM schemes do not fit well into the TCP design space. Mapping from the direct rate control in NUM schemes to a TCP window-control implementation is non-trivial, since source rate is never a simple, rational function of its window size; see an example in Fig. 1.1 even when window sizes of all other sources remained fixed and a simplified network fluid model is assumed [59]. In addition, the NUM rate controller at each source requires the knowledge of aggregate queue lengths (or shadow prices as function of queue lengths) along its route. Obtaining this information for the end nodes in a scalable manner (e.g., without explicit feedback from intermediate nodes) is indeed difficult.

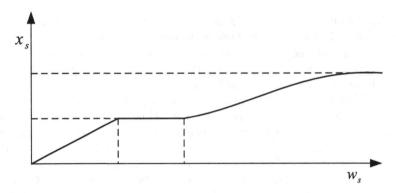

Fig. 1.1 Source rate x_s versus its window size w_s when all other window sizes $w_{s'}$, $\forall s' \neq s$, remain fixed

Without being carefully addressed, these implementation and scalability issues can prevent operating and testing the existing NUM schemes for practical applications.

Most of the heuristics based TCP enhancements lack systematic design procedures and analytically strong performance guarantees. Inspired by the delay based TCP schemes [41–43, 53], the Mo-Walrand scheme and FAST-TCP [2, 59] suggest promising directions towards implementation-friendly yet optimal TCP congestion control. Unlike aggregate queue length, the aggregate queueing delay for a flow can be estimated at the source by e.g., the difference between average round-trip-time (RTT) and the minimum RTT observed [2]. Using this congestion measure preserves the scalability of TCP since explicit feedback from the network is not required. According to this measure, the TCP compatible window controllers were designed and analyzed in Mo-Walrand scheme and FAST-TCP. However, these schemes were only developed for TCP optimization in the wired Internet, where all links have constant and independent capacities. For Internet with coupled wireless links where cross-layer design becomes necessary for global network optimization, there lacks a solid theoretic ground for joint optimization of these TCP congestion control schemes with other-layer protocols and schemes.

Summarizing, the existing network design/optimization approaches are based either on heuristics that are not closely guided by the theory, or on theories that do not sufficiently take into account critical design constraints. As a result, the proposed schemes either cannot provide analytical performance guarantees, or are difficult to be deployed with the current infrastructure, especially for Internet with (coupled) wireless links.

1.3 The Approach

To fill the gap, a tight(-er) integration of the theory and design is clearly called for. To this end, we develop a new *design-space oriented* cross-layer optimization paradigm for wireless Internet applications. Different from the existing theory-guided design methods, the main ideas behind our approach are the use of queueing delays to regulate the protocol interactions across layers, and development of window-based cross-layer optimization algorithms. Almost all Internet traffic uses TCP-based protocols such as the Hypertext Transfer Protocol (HTTP), Simple Mail Transfer Protocol (SMTP), and File Transfer Protocol (FTP), where a window-based congestion control is performed at end users for competing data flows. Such a window-based operation has been long proven to be the key for implementability, reliability and high performance of the Internet protocols since 1988, and all the practical TCP variants built on this window-control mechanism [34–37]. We thus believe that this mechanism constitutes an indispensable part of the TCP design space, and it needs to be explicitly addressed in protocol optimization. Furthermore, a joint design of this TCP window control with other-layer schemes is necessary to global network optimization, especially for wireless applications. It is clearly challenging to accommodate such a cross-layer optimization in the Internet design space that requires distributed, asynchronous and layered operation of network protocols.

To break the ice, we expect queueing delays to play a very important role. When end-to-end queueing delays are adopted as the congestion measure at the source nodes, it can be shown that TCP window-control mechanism amounts to implicit updates of source rates as "primal variables" and queueing delays as "Lagrange dual variables" from optimization-theoretic perspective. Acting as Lagrange multipliers, we propose that queueing delays can further "glue" the TCP window-control with protocols/algorithms at other layers together as the decomposition parts of a global solver for the intended optimization problem. Moreover, since queueing delays can be estimated at the nodes without explicit feedback from the network [1, 53], they are indeed well suited to coordinate the interactions of the Internet protocols consisting of layered, asynchronously, distributed algorithms. Confined to the Internet design space, the proposed cross-layer optimization approach is then to develop non-standard *window-based implicit primal-dual solvers* for underlying optimization problems, and rely on queueing delays to *decompose* these solvers into local algorithms that can be deployed and operated asynchronously at different layers of distributed network nodes.

1.4 Outline of the Book

Our original approach has far-reaching implications and clearly holds promise towards high-performance network protocols/schemes for emerging wireless Internet applications. Development of the proposed cross-layer optimization is, however, challenging. To be compatible with the TCP design space, we need to propose a class of non-standard solvers, where the primal-dual updates must be a mapping of TCP window control implementations. These solvers should also be friendly decomposable such that the resultant jointly optimal schemes can be implemented with the current layered Internet infrastructure. Unlike the classic gradient-type algorithms, there is no general method available for developing these design-space oriented non-standard solvers in the optimization textbooks. To fill the need, this book is devoted to establishing the existence of such solvers and creating a systematic framework for their design and analysis, with a focus on joint design of link-scheduling (at link layer) and congestion control (at transport layer) for wireless Internet applications.

In Chap. 2, we consider an Internet topology with a wired backbone and cellular wireless links. Capitalizing on the proposed approach, we prove the existence of *readily deployable, scalable, yet optimal* joint congestion control and wireless-link scheduling schemes for TCP flows over such a network. Specifically, we show that a queueing-delay based "MaxWeight"-type scheduling should be implemented to make the wireless links coupled in a desired manner. Generalizing the Mo-Walrand scheme [59], we construct a class of window-based QUeueIng-Control (QUIC) algorithms for TCP congestion control. It was shown that these algorithms can be glued with the queueing-delay based MaxWeight-scheduler to constitute window-based implicit primal-dual solvers, for which global convergence to optimal equilibrium can be established in the network fluid model.

In Chap. 3, we generalize our approach to joint optimization of TCP congestion control and distributed link-scheduling for the Internet traffic over ad-hoc wireless links. Building on the window-control oriented primal-dual solution concept, we adopt the QUIC-TCP for flow congestion control. For distributed scheduling of wireless links, we propose a novel carrier sense multiple access (CSMA) algorithm where each link-transmitter employs its queueing delay to properly control its back-off time to access the shared wireless channel. Under this random access scheme, we show that the capacity of the ad-hoc wireless network can be fully utilized. In addition, the wireless links are coupled in a desired manner such that QUIC-TCP congestion control and CSMA scheduling entail an implicit primal-dual solver to the intended optimization problem.

Since the cellular and ad-hoc wireless systems typically operate at different frequency bands in practice, the design and analysis approaches in Chaps. 2 and 3 can be simply combined to provide the solutions for the Internet with both cellular and ad-hoc wireless links. This is delineated in Chap. 4. In Chaps. 2 and 3, the proposed approaches are developed for NUM with a proportionally fair utility function in the network fluid model, where the feedback delays and stochastic traffic/channel dynamics are absent. In Chap. 4, we also generalize our approach to NUM with general concave utility functions. In addition, we outline that generalized Nyquist criterion and stochastic approximation tools can be employed to study the stability and performance of the proposed network schemes in realistic Internet environments with feedback delays and network traffic/channel dynamics, and the proposed approach could be generalized to joint congestion control, routing, and link-scheduling optimization for networks supporting multipath source routing.

Finally, Chap. 5 summarizes the book. As an extension, integration, and re-engineering of the current state-of-art theory-guided network design methods, the proposed novel cross-layer optimization approach is an attempt to bridge the network optimization theory and practical Internet designs. Generalization of the approach is possible to re-develop many existing NUM cross-layer optimization algorithms into high-performance practical schemes that can be deployed and operated over current Internet infrastructure, and embody a paradigm shift in the theory and practice for cross-layer design and optimization of Internet protocols. As a result, the research could impact the *basic theory* in understanding and optimizing the large-scale network and communication systems, and is expected to benefit directly *applications* to next-generation Internet protocol designs.

References

1. C. Jin, D. Wei, S. Low, G. Buhrmaster, J. Bunn, D. Choe, R. Cottrell, J. Doyle, W. Feng, O. Martin, H. Newman, F. Paganini, S. Ravot, S. Singh, FAST TCP: From theory to experiments. IEEE Network **19**(1), 4–11(2005)
2. D. Wei, C. Jin, S. Low, S. Hedge, FAST TCP: motivation, architecture, algorithms, performance. IEEE/ACM Trans. Netw. **14**(6), 1246–1259 (2006)
3. Y. Tian, K. Xu, N. Ansari, TCP in wireless environment: problems and solutions. IEEE Commun. Mag. **43**, 27–32 (2005)
4. H. Balakrishnan, V. Padmanabhan, S. Seshan, R. Katz, A comparison of mechanisms for improving TCP performance over wireless links. IEEE/ACM Trans. Netw. **5**(6), 756–769 (1997)
5. F. Kelly, Fairness and stability of end-to-end congestion control. Eur. J. Control **9**, 159–176 (2003)
6. S. Low, A duality model of TCP and queue management algorithms. IEEE/ACM Trans. Netw. **11**(4), 525–536 (2003)
7. R. Srikant, *The Mathematics of Internet Congestion Control* (Birkhauser, Cambridge, 2004)
8. F. Kelly, A. Maulloo, D. Tan, Rate control in communication networks: shadow prices, proportional fairness and stability. J. Oper. Res. Soc. **49**(3), 237–252 (1998)
9. S. Low, D. Lapsley, Optimization flow control. I: Basic algorithm and convergence. IEEE/ACM Trans. Netw. **7**(6), 861–874 (1999)
10. M. Chiang, Balancing transport and physical layers in wireless multihop networks: jointly optimal congestion control and power control. IEEE J. Sel. Areas Commun. **23**(1), 104–116 (Jan. 2005)
11. J. Lee, M. Chiang, R. Calderbank, Price-based distributed algorithm for optimal rate-reliability tradeoff in network utility maximization. IEEE J. Sel. Areas Commun. **24**(5), 962–976 (2006)
12. L. Chen, S. Low, J. Doyle, Joint congestion control and media access control design for ad hoc wireless networks, *Proceedings of IEEE INFOCOM Conference*, vol. 3, pp. 2212–2222, Miami, FL, 13–17 Mar. 2005
13. X. Wang, K. Kar, Cross-layer rate optimization for proportional fairness in multihop wireless networks with random access. IEEE J. Sel. Areas Commun. **24**(8), 1548–1559 (2006)
14. J. Liu, A. Stoylar, M. Chiang, H. Poor, Queue back-pressure random access in multihop wireless networks: optimality and stability. IEEE Trans. Inf. Theory **55**(9), 4087–4098 (2009)
15. L. Bui, A. Eryilmaz, R. Srikant, X. Wu, Joint asynchronous congestion control and distributed scheduling for multi-hop wireless networks, *Proceedings of IEEE INFOCOM Conference*, Barcelona, Spain, Apr. 2006
16. A. Eryilmaz, R. Srikant, Fair resource allocation in wireless networks using queue-length-based scheduling and congestion control, *Proceedings of IEEE INFOCOM Conference*, vol. 3, pp. 1794–1803, Miami, FL, 13–17 Mar. 2005
17. B. Johansson, P. Soldata, M. Johansson, Mathematical decomposition techniques for distributed cross-layer optimization of data networks. IEEE J. Sel. Areas Commun. **24**(8), 1535–1547 (2006)

18. M. Neely, E. Modiano, C. Rohrs, Dynamic power allocation and routing for time-varying wireless networks. IEEE J. Sel. Areas Commun. **23**(1), 89–103 (2005)

19. L. Xiao, M. Johansson, S. Boyd, Joint routing and resource allocation via dual decomposition. IEEE Trans. Commun. **52**(7), 1136–1144 (2004)

20. L. Chen, S. Low, M. Chiang, J. Doyle, Cross-layer congestion control, routing, and scheduling design in ad hoc wireless networks, *Proceedings of IEEE INFOCOM Conference*, Barcelona, Spain, Apr. 2006

21. A. Eryilmaz, R. Srikant, Joint congestion control, routing and MAC for stability and fairness in wireless networks. IEEE J. Sel. Areas Commun. **24**(8), 1514–1524 (2006)

22. X. Lin, N. Shroff, The impact of imperfect scheduling on cross-layer rate control in wireless networks. IEEE/ACM Trans. Netw. **14**(2), 302–315 (2006)

23. R. Cruz, A. Santhanam, Optimal routing, link scheduling and power control in multi-hop wireless networks, *Proceedings of IEEE INFOCOM Conference*, vol. 1, pp. 702–711, San Francisco, CA, 30 Mar.–4 Apr. 2003

24. M. Chiang, S. Low, A. Calderbank, J. Doyle, Layering as optimization decomposition. Proc. IEEE **95**(1), 255–312 (2007)

25. D. Palomar, M. Chiang, A tutorial on decomposition methods for network utility maximization. IEEE J. Sel. Areas Commun. **24**(8), 1439–1451 (2006)

26. Y. Yi, M. Chiang, Stochastic network utility maximization: a tribute to Kelly's paper published in this journal a decade ago. Eur. Trans. Telecommun. **19**(4), 421–442 (2008)

27. M. Neely, E. Modiano, C. Li, Fairness and optimal stochastic control for heterogenous networks. IEEE/ACM Trans. Netw. **16**(2), 396–409 (2008)

28. Y. Yu, G. Giannakis, Joint congestion control and OFDMA scheduling for hybrid wireline-wireless networks, *Proceedings of IEEE INFOCOM Conference*, pp. 973–981, Anchorage, AK, 6–12 May 2007

29. P. Bender, P. Black, M. Grob, R. Padovani, N. Sindhushyana, A. Viterbi, CDMA/HDR: a bandwidth efficient high speed wireless data service for nomadic users. IEEE Commun. Mag. **38**(7), 70–77 (2000)

30. A. Stolyar, Maximizing queueing network utility subject to stability: greedy primal-dual algorithm. Queueing Syst. **50**, 401–457 (2005)

31. X. Wang, G.B. Giannakis, A.G. Marques, A unified approach to QoS-guaranteed scheduling for channel-adaptive wireless networks. Proc. IEEE **95**(12), 2410–2431 (2007)

32. L. Tassiulas, A. Ephremides, Stability properties of constrained queueing systems and scheduling policies for maximum throughput in multihop radio networks. IEEE Trans. Autom. Control **36**(12), 1936–1948 (1992)

33. X. Wang, N. Gao, Stochastic resource allocation in fading multiple access and broadcast channels. IEEE Trans. Inf. Theory **56**(5), 2382–2391 (2010)

34. M. Allman, V. Paxson, W. Stevens, TCP congestion control, RFC 2581, Apr. 1999

35. V. Jacobson, R. Braden, D. Borman, TCP extensions for high performance, RFC 1323, May 1992

36. W. Stevens, TCP slow start, congestion avoidance, fast retransmit, and fast recovery algorithms, RFC 2001, Jan. 1997

37. S. Floyd, T. Henderson, The NewReno modification to TCP's fast recovery algorithm, RFC 2582, Apr. 1999

38. S. Floyd, High speed TCP for large congestion windows, RFC 3649, Dec. 2003

39. T. Kelly, Scalable TCP: improving performance in highspeed wide area networks. ACM SIGCOMM Comput. Commun. Rev. **33**(2), 83–91 (2003)

40. J. Liu, S. Singh, ATCP: TCP for mobile ad hoc networks. IEEE J. Sel. Areas Commun. **19**(7), 1300–1315 (2001)

41. C. Fu, S. Liew, TCP Veno: TCP enhancement for transmission over wireless access networks. IEEE J. Sel. Areas Commun. **21**(2), 216–228 (2004)

42. S. Mascolo, C. Casetti, M. Gerla, M. Sanadidi, R. Wang, TCP westwood: bandwidth estimation for enhanced transport over wireless links, *Proceedings of ACM MobiCom Conference*, pp. 287–297, July 2001

43. K. Xu, Y. Tian, N. Ansari, TCP-Jersey for wireless IP communications. IEEE J. Sel. Areas Commun. **22**(4), 747–756 (2004)

44. H. Chaskar, T. Lakshman, U. Madhow, TCP over wireless with link level error control: analysis and design methodology. IEEE/ACM Trans. Netw. **7**(5), 605–615 (1999)

45. M. Chan, R. Ramjee, TCP/IP performance over 3G wireless links with rate and delay variation, *Proceedings of ACM MobiCom Conference*, pp. 71–82, Sept. 2002

46. M. Chan, R. Ramjee, Improving TCP/IP performance over third generation wireless networks, *Proceedings of IEEE INFOCOM Conference*, vol. 3, pp. 1893–1904 (2004)

47. S. ElRakabawy, A. Klemm, C. Lindemann, TCP with active pacing for multihop wireless networks, *Proceedings of ACM MobiHoc Conference*, pp. 288–299, May 2005

48. L. Andrew, S. Hanly, R. Mukhtar, Active queue management for fair resource allocation in wireless networks. IEEE Trans. Mob. Comput. **7**(2), 231–246 (2008)

49. T. Goff, J. Moronski, D. Phatak, V. Gupta, Freeze-TCP: a true end-to-end TCP enhancement mechanism for mobile environments, *Proceedings of ACM MobiHoc Conference*, vol. 3, pp. 1537–1545, Mar. 2000

50. L. Kalampoukas, A. Varma, K. Ramakrishnan, Explicit window adoption: a method to enhance TCP performance. IEEE/ACM Trans. Netw. **10**, 338–350 (2002)

51. S. Spring, M. Chesire, M. Berryman, V. Sahasranaman, T. Anderson, B. Bershad, Receiver-based management of low bandwidth access links, *Proceedings of IEEE INFOCOM Conference*, pp. 245–254 (2000)

52. F. Paganini, Z. Wang, J. Doyle, S. Low, Congestion control for high performance, stability and fairness in general networks. IEEE/ACM Trans. Netw. **13**(1), 43–56 (2005)

53. L. Brakmo, L. Peterson, TCP Vegas: end-to-end congestion avoidance on a global internet. IEEE J. Sel. Areas Commun. **13**(8), 1465–1480 (1995)

54. A. Kuzmanovic, E. Knightly, TCP-LP: a distributed algorithm for low priority data transfer, *Proceedings of IEEE INFOCOM Conference*, pp. 1691–1701 (2003)

55. L. Xu, K. Harfoush, I. Rhee, Binary increase congestion control (BIC) for fast long-distance networks, *Proceedings of IEEE INFOCOM Conference*, pp. 2514–2524 (2004)

56. S. Jin, L. Guo, I. Matta, A. Bestavros, A spectrum of TCP-friendly window-based congestion control algorithms. IEEE/ACM Trans. Netw. **11**(3), 341–355 (2003)

57. S. Ha, I. Rhee, L. Xu, CUBIC: a new TCP-friendly high-speed TCP variant. ACM SIGOPS Oper. Syst. Rev. **42**(5), 64–74 (2008)

58. K. Tan, J. Song, Q. Zhang, M. Sridharan, Compound TCP: a scalable and TCP-friendly congestion control for high-speed networks, *Proceedings of PFLDnet* (2006)

59. J. Mo, J. Walrand, Fair end-to-end window-based congestion control. IEEE/ACM Trans. Netw. **8**(5), 556–567 (2000)

Chapter 2
Joint Congestion Control and Link Scheduling for Internet with Cellular Wireless Links

To provide ubiquitous data services, the Internet has extended from its wired backbone to include an increasing number of wireless links to serve mobile users. For simplicity, we first consider an Internet with a wired backbone and an access point that provides one-hop communication for wireless devices in a single cell; see Fig. 2.1. Generalization to multi-cell case is straightforward. As delineated in the last chapter, joint design/optimization of the network protocols for Internet applications over this hybrid wired-wireless network is challenging. To this end, we propose to extend the *window-based implicit primal-dual solver* approach [1, 2] to cross-layer optimization, and rely on queueing delays to *decompose* such a global solver into local algorithms that can be deployed and operated asynchronously at different layers of distributed network nodes. Capitalizing on this idea, we investigate joint TCP congestion control (at transport layer) and wireless-link scheduling (at link layer) schemes for wireless applications over such a network.

2.1 Network Model and Problem Formulation

A logical data link is a transmitter-receiver pair. The set of logical links $L = L_f \cup L_w$ in the network consists of a wired link set L_f and a wireless link set L_w. Any wired link $l \in L_f$ is assumed to have a constant and independent link capacity c_l. Let r_l denote the capacity of wireless link l and define a vector $r := \{r_l, \; \forall l \in L_w\}$. In wireless standards for cellular systems, such as IEEE 802.16, UMTS HSDPA and CDMA2000 EV-DO, link-layer schemes including error correction coding, fast link (rate/power) adaptation, and hybrid Automatic Repeat-reQuest (ARQ) protocol are incorporated to provide reliable transmissions over wireless links. Hence, we assume that the packet loss over wireless links can be sufficiently masked; yet the link capacities r_l are dynamic and coupled due to the rate adaptation and broadcast nature of the wireless medium. All wireless links are subject to random fading. Per fading realization h, the largest rate set for the coupled wireless links is dictated by

X. Wang, *Scheduling and Congestion Control for Wireless Internet*,
SpringerBriefs in Electrical and Computer Engineering,
DOI: 10.1007/978-1-4614-8420-2_2, © The Author(s) 2014

Fig. 2.1 Internet with a wired
backbone and cellular wireless
links

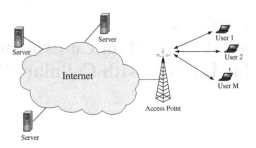

a region $\mathcal{R}(h)$, which is assumed closed and convex. On average we have $\bar{r} =:$ $\mathbb{E}_h[r(h)] \in \bar{\mathcal{R}}$ where $\mathbb{E}_h$ denotes the expectation over fading distribution.

For the closed and convex $\bar{\mathcal{R}}$, we can easily show that there exists a convex function $f(\bar{r})$ such that we can express $\bar{\mathcal{R}} = \{\bar{r} \geq 0 \mid f(\bar{r}) \leq 0\}$ [3]. Consider a smooth capacity region, for which we can find a strictly convex and twice differentiable $f(\bar{r})$ to define its boundary. We have such a region when continuous rate/power adaptation, or approximately a sufficiently large number of rate/power levels (e.g., 30 levels in UMTS HSPDA), are supported by the wireless links.

Suppose that the network is shared by a set of unicast flows from sources $s \in S$. With a given routing table, each flow s goes through multiple wired and wireless links. Let $L(s) \subseteq L$ be the set of links that flow source s uses, and $S(l) \subseteq S$ the set of flows that use link l. Let x_s denote the sending rate of source s. With a weight vector $p := \{p_s, \forall s\}$, adopt a weighted proportionally-fair utility function $p_s \log x_s$ for each flow [2, 4]. We consider the aggregate utility maximization subject to (s. t.) link capacity constraints:

$$\max \quad \sum_{s \in S} p_s \log x_s$$

$$\text{s. t. (C1): } \sum_{s \in S(l)} x_s \leq c_l, \ \forall l \in L_f$$

$$\text{(C2): } \sum_{s \in S(l)} x_s \leq \bar{r}_l, \ \forall l \in L_w \qquad (2.1)$$

$$\text{(C3): } \bar{r} \in \bar{\mathcal{R}}, \quad x_s \geq 0, \ \forall s \in S.$$

The utility function in (2.1) is used as an example for elaboration purpose only; it is in fact the $(p, 1)$-proportionally fair function—a special case of the (p, α)-proportionally fair functions. These utility functions are of interest, since it was shown that maximization of them can lead to an efficient and fair network equilibrium that does not penalize the flows with large propagation delays [1, 2, 4].

2.2 Algorithm Development and Analysis

The simple NUM problem (2.1) is to solve a jointly optimal congestion control and wireless-link scheduling scheme for a hybrid wired-wireless network. With queue lengths playing the role of Lagrange multipliers, decomposable gradient-type primal-dual solvers can be readily derived with the classic NUM paradigm [5–8]. Aiming at *readily deployable, scalable yet provably optimal* network schemes, we next develop a novel window-control oriented cross-layer optimization approach for this "old" problem.

Let $\{x^*, \bar{r}^*\}$ denote the optimal solutions for the convex program (2.1) and $\lambda^* := \{\lambda_l^*, \forall l \in L\}$ the optimal dual vector for its dual problem. Then the sufficient and necessary Karush-Kuhn-Tucker (KKT) optimality conditions imply that $\{x^*, \bar{r}^*\}$ and λ^* satisfy [3]:

$$\frac{p_s}{x_s^*} = \lambda^{s*} := \sum_{l \in L(s)} \lambda_l^*, \quad \forall s \in S \tag{2.2}$$

$$\lambda_l^*\left(c_l - \sum_{s \in S(l)} x_s^*\right) = 0, \quad \sum_{s \in S(l)} x_s^* \le c_l, \quad \forall l \in L_f \tag{2.3}$$

$$\lambda_l^*\left(\bar{r}_l^* - \sum_{s \in S(l)} x_s^*\right) = 0, \quad \sum_{s \in S(l)} x_s^* \le \bar{r}_l^*, \quad \forall l \in L_w \tag{2.4}$$

$$\bar{r}^* = \arg\max_{\bar{r} \in \mathscr{R}} \sum_{l \in L_w} \lambda_l^* \bar{r}_l, \quad x^* \ge 0, \quad \lambda^* \ge 0. \tag{2.5}$$

2.2.1 Joint Design in Network Fluid Model

To guide our design, consider a (simplified) fluid model of network where packets are infinitely divisible and small. In this model, let $w := \{w_s, \forall s\}$ collect the window sizes for all sources $s \in S$, $q := \{q_l, \forall l\}$ collect the round-trip queueing delays for all links $l \in L$, and $d := \{d_s, \forall s\}$ collect the fixed round-trip propagation (plus processing) delays for all sources. Upon defining the aggregate queueing delays $q^s := \sum_{l \in L(s)} q_l$ along the flow routes, we have the following relationships for source rates x_s, window sizes w_s, and queueing delays q_l [1]:

$$x_s(d_s + q^s) = w_s, \quad \forall s \in S \tag{2.6}$$

$$q_l\left(c_l - \sum_{s \in S(l)} x_s\right) = 0, \quad \sum_{s \in S(l)} x_s \le c_l, \quad \forall l \in L_f \tag{2.7}$$

$$q_l\left(\bar{r}_l - \sum_{s \in S(l)} x_s\right) = 0, \quad \sum_{s \in S(l)} x_s \le \bar{r}_l, \quad \forall l \in L_w \tag{2.8}$$

$$\bar{r} \in \bar{\mathscr{R}}, \quad x \geq 0, \quad q \geq 0. \tag{2.9}$$

Here (2.6) follows from that the source rate x_s is equal to the window size w_s divided by the total round-trip delay $d_s + q^s$ of flow s; (2.7) and (2.8) are implied by the link capacity constraints and the fact that if the aggregate rate through a link l is less than its capacity, the queueing delay at this link is equal to zero since packets are infinitely divisible and small.

The fluid model identities (2.6)–(2.9) in fact define a mapping from the window sizes to the source rates and queueing delays: $F : w \rightarrow (x, q)$. Specifically, construct a convex program:

$$\max_{x,\bar{r}} \sum_{s \in S} (w_s \log x_s - d_s x_s), \quad \text{s. t.} \quad \text{(C1)–(C3) of (2.1).} \tag{2.10}$$

Careful examination can tell that (2.6)–(2.9) correspond exactly to the KKT conditions of (2.10), with q playing the role of Lagrange multipliers associated with the constraints (C1) and (C2) [3]. Therefore, the mapping F is defined by the solutions of (2.10) and its dual problem. Through this mapping, the TCP congestion control in the fluid model is to adjust the window size w_s in order to control the source-rate x, wireless-link capacity $\bar{r}$ and queueing delay q towards a desired network equilibrium that can yield the optimal solution for the NUM problem (2.1).

To this end, comparing the KKT conditions (2.2)–(2.5) and fluid model identities (2.6)–(2.9) naturally implies that a queueing-delay based link scheduler should be implemented at the wireless access point per fading state h:

$$r^*(h) = \arg\max_{r \in \mathscr{R}(h)} \sum_{l \in L_w} q_l r_l, \quad \text{such that} \quad \bar{r}^* := \mathbb{E}_h\left[r^*(h)\right] = \arg\max_{r \in \bar{\mathscr{R}}} \sum_{l \in L_w} q_l r_l. \tag{2.11}$$

This scheduler is indeed a MaxWeight-type scheduler similar to those in [7–9]. However, the queueing delays (instead of queue lengths) are employed as the weights for user rates, as queueing delays play the role of Lagrange multipliers in our design.

On the other hand, since $x_s q^s = w_s - x_s d_s$ from (2.6) while $p_s = x_s^* \lambda^{s*}$ in (2.2), the difference $v_s := w_s - x_s d_s - p_s$ could serve as a distance between the current operating point and the optimal network equilibrium. Using this distance, we generalize the Mo-Walrand scheme [1] to propose a class of window updates following the ordinary differential equations (ODEs):

$$\frac{d}{dt} w_s(t) = -\frac{d_s}{\bar{d}_s} w_s^{-2\rho+1} v_s, \quad \forall s \tag{2.12}$$

where the total round-trip delay $\bar{d}_s := d_s + q^s$, and ρ is a constant between 0 and 1. Since the window adjustment (2.12) is to control the queueing of the flows, we call it a QUIC (QUeueIng Control) algorithm parameterized by a $\rho \in [0, 1]$. This class of algorithms specializes to the Mo-Walrand scheme [1]: $\frac{d}{dt} w_s(t) = -\frac{d_s}{\bar{d}_s} \frac{v_s}{w_s}$ as $\rho = 1$.

With $\rho = 1/2$, the proposed window update becomes $\frac{d}{dt}w_s(t) = -\frac{d_s}{d_s}v_s$, which is similar to that in FAST-TCP [2]: $\frac{d}{dt}w_s(t) = -v_s$.

In the equilibrium of window update (2.12), we clearly have $v := \{v_s, \forall s\} = 0$. Let w^* denote the window-size vector with this equilibrium, and x^*, $\bar{r}^*$ and q^* the corresponding source-rate, wireless-link capacity, and queueing-delay vectors. We can establish that:

Theorem 2.1 *For the proposed joint design, there is a unique window-size vector w^* such that $v = 0$, and the rate vector $x^* = x(w^*)$ is the p-weighted proportionally-fair rate vector, i.e., it maximizes (2.1).*

Proof See Appendix 2A.

Theorem 2.1 states that by pushing $v_s = 0$, $\forall s$, we can obtain the optimal w^* that leads to the optimal source rate vector x^* for (2.1); hence, $v_s = 0$ can serve as decoupled optimality criteria for the sources since v_s can be evaluated using local w_s, d_s, and x_s. Under the link-scheduler (2.11), window update (2.12) is a negative feedback system towards w^*. When w_s is large such that $v_s = w_s - x_s d_s - p_s > 0$, we have a negative dw_s/dt to decrease the window size so that v_s is decreased towards zero; and vice verse.

2.2.2 Convergence Analysis

Having shown that our joint design has a unique equilibrium per Theorem 2.1, we next analytically establish that the proposed schemes globally converge to this optimal operating point from any initial state. To this end, we first study the coupling of wireless links under the proposed scheduler. Recall that we model the capacity region $\bar{\mathscr{R}} = \{\bar{r} \mid \bar{r} \geq 0, \ f(\bar{r}) \leq 0\}$ where $f(\bar{r})$ is strictly convex and twice differentiable. Let $q_w := \{q_l, \forall l \in L_w\}$ and ∇f denote the gradient of f. For the queueing-delay based scheduler (2.11), we can establish that:

Lemma 2.1 *Given a strictly convex $f(\bar{r})$, we have $\bar{r}^* = \arg\max_{\bar{r} \in \bar{\mathscr{R}}} \sum_{l \in L_w} q_l \bar{r}_l$ if and only if $f(\bar{r}^*) = 0$ and $\theta \nabla f(\bar{r}^*) = q_w$ for a certain constant $\theta > 0$.*

Proof See Appendix 2B.

As illustrated in Fig. 2.2, Lemma 2.1 simply indicates that the rate vector $\bar{r}^*$ resulted from the queueing-delay based scheduling should reside on the boundary of $\bar{\mathscr{R}}$ (i.e., $f(\bar{r}^*) = 0$) and the gradient of f at this point has the same direction with the queueing-delay vector q_w (i.e., $\theta \nabla f(\bar{r}^*) = q_w$). The resultant $\bar{r}^*$ on the boundary of $\bar{\mathscr{R}}$ guarantees that the proposed scheduler fully utilizes the capacity of wireless links. On the other hand, $\nabla f(\bar{r}^*)$ having the same direction with q_w can render the wireless links to be coupled in a "desired" manner. Let $J_{\bar{r}^* | q_w} := \{\frac{\partial \bar{r}_l^*}{\partial q_n}, \forall l, n \in L_w\}$

Fig. 2.2 Geometric interpre-
tation of Lemma 2.1

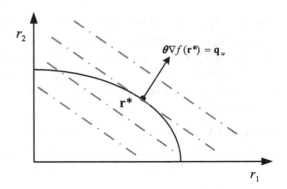

denote the Jacobian matrix of vector $\bar{r}^*$ with respect to q_w. From Lemma 2.1, we can
further prove:

Lemma 2.2 *Given a strictly convex and twice differentiable $f(\bar{r})$, the Jacobian
$J_{\bar{r}^*|q_w}$ is positive definite under the scheduler (2.11).*

Proof See Appendix 2B.

Lemma 2.2 shows that by using queueing delays to make scheduling decisions in
(2.11), the wireless links are coupled in a way such that $J_{\bar{r}^*|q_w}$ is positive definite.
Using this nice property, we can establish the following global stability result:

Theorem 2.2 *Under the scheduler (2.11), the window-based control (2.12) with
$0 \le \rho \le 1$ globally converges to its unique equilibrium w^*, and the source-rate
vector converges to x^*.*

Proof See Appendix 2C.

Theorem 2.2 is a generalization of [1, Theorem 5], which holds only for the QUIC-
TCP scheme (2.12) with $\rho = 1$ (i.e. Mo-Walrand scheme) in *wired* networks. Here
we prove that the global stability of QUIC-TCP with $0 \le \rho \le 1$ can be established for
general wired-wireless networks, when a queueing-delay based MaxWeight strategy
is employed for wireless-link scheduling. The proof is based on the classic Lyapunov
method, and a key for the generalization to wireless setup is the positive definiteness
of $J_{\bar{r}^*(q)|q_w}$ under the proposed scheduler (2.11). Specifically, we show that $Y(w) =
(1/2) \sum_{s \in S}(v_s/(w_s)^\rho)^2$ is a Lyapunov function for (2.11) and (2.12); consequently,
global convergence of the proposed schemes to its unique equilibrium readily follows.

Note that the proposed QUIC-TCP scheme becomes $\frac{d}{dt}w_s(t) = -\frac{d_s}{\bar{d}_s}v_s$ with
$\rho = 1/2$, which is close to the window control in FAST-TCP: $\frac{d}{dt}w_s(t) = -v_s$
[2], especially when all flows have similar $d_s/\bar{d}_s$ values. The FAST-TCP was proven
to be locally stable in general *wired* networks, or globally stable for a single-link
wired network; yet, the global stability of FAST-TCP for general networks is still
an open problem. Different from the FAST-TCP, global stability/convergence of the

QUIC-TCP is proven in Theorem 2.2 for general networks with wired and coupled wireless links.

2.2.3 Practical Algorithm Development

Using the insights provided by the fluid model based design, we next develop the readily deployable congestion control and link scheduling algorithms for practical systems.

Congestion control: As with FAST-TCP, the proposed congestion control (at transport-layer) is divided into four functionally components: estimation, window control, data control, and burstiness control components [2]. Whenever an in-order acknowledge (ACK) packet is received, the estimation component per flow s computes the current round-trip time RTT_s and updates the values of $AvgRTT_s$ and $BaseRTT_s$, The value of $BaseRTT_s$ is obtained as the minimum RTT observed so far by the source to approximate d_s. Given the current RTT_s, the source also updates the average total round-trip delay $AvgRTT_s$ via a low-pass filter; e.g.,

$$AvgRTT_s \leftarrow \frac{255}{256} \times AvgRTT_s + \frac{1}{256} \times RTT_s,$$

to estimate $\bar{d}_s$. Note that $v_s = w_s - x_s d_s - p_s = w_s - \frac{w_s}{d_s} d_s - p_s$. With $BaseRTT_s$ and $AvgRTT_s$ playing the role of d_s and $\bar{d}_s$, respectively, we then propose that the window control component adjusts the transmission window size w_s [cf. (2.12)]:

$$w_s \leftarrow w_s - \kappa \frac{BaseRTT_s}{AvgRTT_s} w_s^{-2\rho+1} \left(w_s - \frac{BaseRTT_s}{AvgRTT_s} w_s - p_s \right) \qquad (2.13)$$

where κ is a positive stepsize. The data control component then determines which packets to be transmitted and the burstiness control component determines when to transmit these packets [2]. When an in-order ACK is not received, the slow start and/or fast recovery schemes in standard TCP are preserved to deal with transmission time-outs or duplicate ACKs.

Wireless-link scheduling: To estimate its queueing delay at a scheduling time t, the access point reads the queue length $QueLEN_l(t)$ and calculates the average rate $AveR_l(t)$ using a similar low-pass filter for each wireless link. Implied by the Little's law, the average queueing delay is given by the ratio of average queue length and average rate. Based on the fluid limit argument [6], the "stochastic" delay $\frac{QueLEN_l(t)}{AveR_l(t)}$, i.e., current queue length divided by the average rate, could play the role of average delay q_l in the deterministic fluid model. The link-scheduler at the access point is then to select a strategy such that [cf. (2.11)]:

$$r^*(t) = \arg \max_{r \in \mathscr{R}(h(t))} \sum_{l \in L_w} \frac{QueLEN_l(t)}{AveR_l(t)} r_l. \tag{2.14}$$

For a shared slotted downlink or uplink in e.g. CDMA 2000 EV-DO and IEEE 802.16 WiMax, suppose that $R_l[n]$ is the rate capacity that link l can currently support per slot n, $\forall l \in L_w$. Then the access point simply schedules the link with the highest $\frac{QueLEN_l[n]}{AveR_l[n]} R_l[n]$ per slot n. Such a scheduler becomes a modified largest-weighted-delay-first (M-LWDF) scheduler which can maintain network stability whenever possible [10].

Remark 2.1 Although the proposed schemes are resulted from a cross-layer design, they can be implemented over the current layered Internet infrastructure. The congestion controller (2.13) preserves the distributed end-to-end mechanism of TCP without message passing from the intermediate nodes, whereas the wireless-link scheduler (2.14) can be easily modified from the existing scheduling policy at the access points of one-hop (cellular) wireless networks. Based on local observations, these schemes entail an *implicit primal-dual* update of $\{x, \bar{r}, q\}$ to solve (2.1). This is different from the gradient-type *explicit primal-dual schemes* in the queue-length based NUM paradigm [6–8], which are difficult to be deployed over Internet infrastructure. Also different from the heuristic schemes in e.g., [11–14], the proposed schemes are *provably optimal* in the network fluid model, thus provide analytical performance guarantees.

2.3 Summary

We prove the existence of readily deployable, scalable, yet optimal joint TCP and wireless-link scheduling schemes for Internet over coupled wireless links, *without explicit message passing*. The success hinges on the tight integration of theory and design. In our theory-guided design approach, analysis is actually done *at* the design time, *not after*. Having clarified the design space and the optimization goal, we perform a Lyapunov function based stability analysis in the simplified network fluid model to guide the development of the non-standard window-based implicit solvers. There is no general technique for constructing Lyapunov functions. With queueing delays playing the role of Lagrange multipliers, we rely on the KKT optimality conditions to identify the desired equilibrium and subsequently propose a quadratic candidate function $Y(w) = (1/2) \sum_{s \in S} (v_s/(w_s)^\rho)^2$. By checking the first time-derivative $\frac{d}{dt} Y(w(t))$, we are then able to derive the window-update ODEs (2.12) and queueing-delay based scheduler (2.11) to render $Y(w)$ a Lyapunov function for the desired equilibrium. The proposed schemes (2.13) and (2.14) are inferred by the fluid model identities (2.11) and (2.12). With design-space explicitly taken into account in the integral design and analysis process, the proposed schemes can be readily deployed and tested over the current infrastructure.

Appendix 2A: Proof of Theorem 2.1

At the equilibrium $v = 0$, we have $w_s^* - x_s^* d_s - p_s = 0$; and it follows from (2.6) that $w_s^* = x_s^*(d_s + q^{s*})$. Hence, we readily have $p_s = x_s^* q^{s*}$, $\forall s$. This is exactly the KKT condition (2.2) if we let $\lambda^* \equiv q^*$. With this equivalence mapping, the fluid model identities (2.7)–(2.9) and the proposed scheduler (2.11) for $\{x^*, \bar{r}^*, q^*\}$ also become the KKT conditions (2.3)–(2.5). Since $x^*, \bar{r}^*$ and q^* satisfy the sufficient and necessary KKT optimality conditions, they are the optimal solutions of (2.1) and its dual problem. Furthermore, we can show that w^* always exists and it is unique due to the existence and uniqueness of x^* for (2.1), as well as the uniqueness of mapping $w^* \to x^*$.

Appendix 2B: Proof of Lemmas

Proof of Lemma 2.1 Since $f(\bar{r})$ is strictly convex, we have

$$f(\bar{r}) > f(\bar{r}^*) + [\nabla f(\bar{r}^*)]^T (\bar{r} - \bar{r}^*), \quad \forall \bar{r} \neq \bar{r}^*.$$

This implies that: $[\nabla f(\bar{r}^*)]^T (\bar{r}^* - \bar{r}) > f(\bar{r}^*) - f(\bar{r}), \quad \forall \bar{r} \neq \bar{r}^*.$
  If we have $f(\bar{r}^*) = 0$ and $\theta \nabla f(\bar{r}^*) = q_w$, then it readily follows that

$$q_w^T(\bar{r}^* - \bar{r}) > \theta(f(\bar{r}^*) - f(\bar{r})) \geq 0, \quad \forall \bar{r} \neq \bar{r}^*, \ \bar{r} \in \bar{\mathcal{R}}$$

since $f(\bar{r}) \leq 0 = f(\bar{r}^*), \forall \bar{r} \in \bar{\mathcal{R}}$. This means exactly: $\bar{r}^* = \arg\max_{\bar{r} \in \bar{\mathcal{R}}} \sum_{l \in L_w} q_l \bar{r}_l$.
  On the other hand, for a strictly convex $f(\bar{r})$, its gradient $\nabla f(\bar{r})$ is continuous and can take any value in the non-negative orthant. Therefore, there always exists a $\bar{r}^*$ such that it satisfies $f(\bar{r}^*) = 0$ and $\theta \nabla f(\bar{r}^*) = q_w$ for any $q_w \geq 0$. The converse thus also holds. $\qquad\square$

Proof of Lemma 2.2 Differentiating both sides of $\theta \nabla f(\bar{r}^*) = q_w$, we have that: $\theta \nabla^2 f(\bar{r}^*) J_{\bar{r}^*|q_w} = I$, where I denotes the $(|L_w| \times |L_w|)$ identity matrix. This implies that: $J_{\bar{r}^*|q_w} = (1/\theta)[\nabla^2 f(\bar{r}^*)]^{-1}$. Since $\nabla^2 f(\bar{r}^*)$ is positive definite, so is $J_{\bar{r}^*|q_w}$. $\square$

Appendix 2C: Proof of Theorem 2.2

For convenience, let $c := \{c_l, \forall l \in L_f\}$ and rewrite problem (2.1) as:

$$\max_{s \in S} \sum p_s \log x_s \quad \text{s.t.} \ A_f x \leq c, \ A_w x \leq \bar{r}, \ \bar{r} \in \bar{\mathcal{R}}, \ x \geq 0 \qquad (2.15)$$

where the routing matrix $A := [A_f^T, \; A_w^T]^T$ with its (l, s)th entry $A_{ls} = 1$ if $s \in S(l)$ and $A_{ls} = 0$ otherwise. A link is a "bottleneck link" if the aggregate rate through it is equal to its capacity. Note that for a given w, the set of bottleneck links contains *either no or all wireless links*. If the bottleneck set contains only part of the wireless links, we have some $q_l > 0$ and some $q_l = 0$ for $l \in L_w$. According to the scheduling strategy (2.11), the links with $q_l = 0$ must have $\bar{r}_l^* = 0$; hence, the rates x_s for the sources that use these links must be zero. This is impossible since the source-rate vector for a given w must be the solution to (2.10), while a rate vector with some $x_s = 0$ makes the objective function of (2.10) become $-\infty$ and thus cannot be the solution for (2.10).

For a given w, let $\mathscr{B}$ denote the set of bottleneck links. Denote A_B as the sub-matrix of A obtained by keeping only the rows that correspond to bottleneck links, and q_B and c_B the corresponding sub-vectors of q and c for bottleneck links. Define the diagonal matrices $X := \text{diag}(x)$, $W := \text{diag}(w)$, $D := \text{diag}(d)$, and $\bar{D} := \text{diag}(\bar{d})$ where $\bar{d} := \{\bar{d}_s, \forall s\}$. Recalling that all non-bottleneck links have zero queueing delays, we can rewrite (2.6) in the matrix form:

$$X(A_B^T q_B + d) = w. \tag{2.16}$$

Suppose first that w is an "interior" point such that the set B remains unchanged within a neighborhood, and $x(w)$ and $q^s(w)$ become differentiable. Note that $A_B^T q_B + d = \bar{d}$. Differentiating both sides of (2.16) with respect to w yield:

$$\bar{D} J_{x|w} + X A_B^T J_{q_B|w} = I. \tag{2.17}$$

Multiplying both sides of (2.17) by $A_B \bar{D}^{-1}$, we further have:

$$A_B J_{x|w} + A_B \bar{D}^{-1} X A_B^T J_{q_B|w} = A_B \bar{D}^{-1}. \tag{2.18}$$

Consider the case that the bottleneck set contains all wireless links. Partition the bottleneck-only routing matrix A_B and queueing delay vector q_B into two parts:

$$A_B = \begin{bmatrix} A_{f,B} \\ A_w \end{bmatrix}, \quad q_B = \begin{bmatrix} q_{f,B} \\ q_w \end{bmatrix}$$

where subscripts $_{f,B}$ and $_w$ denote the parts related to the wired and wireless bottleneck links.

For the wired bottlenecks, it holds $A_{f,B} x = c_B$; thus, $A_{f,B} J_{x|w} = 0$. On the other hand, it holds for wireless bottlenecks: $A_w x = \bar{r}^*$. This implies: $A_w J_{x|w} = [0 \; J_{\bar{r}^*|q_w}] J_{q_B|w}$. Overall, we have:

$$A_B J_{x|w} = \begin{bmatrix} A_{f,B} \\ A_w \end{bmatrix} J_{x|w} = \begin{bmatrix} 0 & 0 \\ 0 & J_{\bar{r}^*|q_w} \end{bmatrix} J_{q_B|w} := N J_{q_B|w}$$

Since $J_{\bar{r}*|q_w}$ is positive definite from Lemma 2.2, the matrix N is positive semi-definite. From (2.18), we have: $(N + A_B\bar{D}^{-1}XA_B^T)J_{q_B|w} = A_B\bar{D}^{-1}$; hence, $J_{q_B|w} = (N + A_B\bar{D}^{-1}XA_B^T)^{-1}A_B\bar{D}^{-1}$.

Substituting the latter into (2.17), we further obtain:

$$J_{x|w} = \bar{D}^{-1}(I - XA_B^T(N + A_B\bar{D}^{-1}XA_B^T)^{-1}A_B\bar{D}^{-1}).$$

Define the matrix

$$M := A_B^T(N + A_B\bar{D}^{-1}XA_B^T)^{-1}A_B. \tag{2.19}$$

Since matrix N is positive semi-definite and matrices $\bar{D}^{-1}$, X are diagonal matrices with positive diagonal entries, matrix M is positive semi-definite; and $J_{x|w} = \bar{D}^{-1}(I - XM\bar{D}^{-1})$.

For the QUIC algorithms (2.12), consider a function $Y(w) = \frac{1}{2}\sum_{s\in S}\left(\frac{v_s}{(w_s)^\rho}\right)^2$. At an interior w, we have:

$$\frac{d}{dt}Y(w(t)) = \sum_s\left(\frac{\partial Y}{\partial w_s}\frac{dw_s(t)}{dt}\right)$$

$$= \sum_s\left[\left(\sum_{s'}\left(\frac{v_{s'}}{w_{s'}^\rho}\frac{\partial(v_{s'}/w_{s'}^\rho)}{\partial w_s}\right)\right)\frac{dw_s(t)}{dt}\right]$$

$$= -v^T W^{-\rho}[W^{-\rho}J_{v|w} - \rho W^{-\rho-1}V]D\bar{D}^{-1}W^{-2\rho+1}v$$

$$= -v^T[W^{-2\rho}(I - DJ_{x|w}) - \rho W^{-2\rho-1}(W - D\bar{D}^{-1}W - P)]D\bar{D}^{-1}W^{-2\rho+1}v$$

$$= -v^T[W^{-2\rho}(I - D\bar{D}^{-1}(I - XM\bar{D}^{-1})) - \rho W^{-2\rho-1}(W - D\bar{D}^{-1}W - P)]$$
$$\times D\bar{D}^{-1}W^{-2\rho+1}v$$

$$= -v^T[(1-\rho)W^{-2\rho}(I - D\bar{D}^{-1})D\bar{D}^{-1}W^{-2\rho+1} + \rho W^{-2\rho-1}PD\bar{D}^{-1}W^{-2\rho+1}$$
$$+ W^{-2\rho+1}D\bar{D}^{-2}M\bar{D}^{-2}DW^{-2\rho+1}]v \tag{2.20}$$

where for convenience we define diagonal matrices $V := \mathrm{diag}(v)$ and $P := \mathrm{diag}(p)$. Since $(I - D\bar{D}^{-1})$ is a diagonal matrix with nonnegative entries, M is positive semi-definite, and all W, D, $\bar{D}$ and P are diagonal matrices with positive diagonal entries, it is easy to see that the whole matrix inside the square bracket of (2.20) is positive definite for $0 \le \rho \le 1$.

In the case that the bottleneck set contains no wireless links, we have $N = 0$. The above proof readily follows too. This implies that $dY(w(t))/dt < 0$ at all interior points, unless $v = 0$.

For the "boundary" (i.e., non-interior) points, we can extend the definition of $J_{x|w}$ as a function of direction d since the right-hand directional derivative of $x(w)$ is always well defined for arbitrary directions. With this extension, we can follow the similar lines in the proof of [1, Theorem 5] to argue that $Y(w(t))$ is also strictly decreasing in t at boundary points unless $v = 0$.

Since $Y(w(t))$ is a nonnegative Lyapunov function with globally negative time derivative, the unique equilibrium $v = 0$ of the system (2.12) is thus globally stable asymptotically; i.e., (2.12) globally converges to its unique equilibrium w^* and consequently the source rates converge to x^*.

References

1. J. Mo, J. Walrand, Fair end-to-end window-based congestion control. IEEE/ACM Trans. Netw. **8**(5), 556–567 (2000)
2. D. Wei, C. Jin, S. Low, S. Hedge, FAST TCP: motivation, architecture, algorithms, performance. IEEE/ACM Trans. Netw. **14**(6), 1246–1259 (2006)
3. S. Boyd, L. Vandenberghe, *Convex Optimization* (Cambridge University Press, Cambridge, 2004)
4. L. Brakmo, L. Peterson, TCP Vegas: end-to-end congestion avoidance on a global internet. IEEE J. Sel. Areas Commun. **13**(8), 1465–1480 (1995)
5. M. Chiang, S. Low, A. Calderbank, J. Doyle, Layering as optimization decomposition. Proc. IEEE **95**(1), 255–312 (2007)
6. A. Stolyar, Maximizing queueing network utility subject to stability: greedy primal-dual algorithm. Queueing Syst. **50**, 401–457 (2005)
7. Y. Yu, G. Giannakis, Joint congestion control and OFDMA scheduling for hybrid wireline-wireless networks, *Proceedings of IEEE INFOCOM Conference*, pp. 973–981, Anchorage, AK, 6–12 May 2007
8. X. Lin, N. Shroff, The impact of imperfect scheduling on cross-layer rate control in wireless networks. IEEE/ACM Trans. Netw. **14**(2), 302–315 (2006)
9. M. Neely, E. Modiano, C. Li, Fairness and optimal stochastic control for heterogenous networks. IEEE/ACM Trans. Netw. **16**(2), 396–409 (2008)
10. M. Andrews, K. Kumaran, K. Ramanan, A. Stolyar, P. Whiting, R. Vijayakumar, Providing quality of service over a shared wireless link. IEEE Commun. Mag. **39**, 150–154 (2001)
11. C. Fu, S. Liew, TCP Veno: TCP enhancement for transmission over wireless access networks. IEEE J. Sel. Areas Commun. **21**(2), 216–228 (2004)
12. K. Xu, Y. Tian, N. Ansari, TCP-Jersey for wireless IP communications. IEEE J. Sel. Areas Commun. **22**(4), 747–756 (2004)
13. S. Ha, I. Rhee, L. Xu, CUBIC: a new TCP-friendly high-speed TCP variant. ACM SIGOPS Oper. Syst. Rev. **42**(5), 64–74 (2008)
14. K. Tan, J. Song, Q. Zhang, M. Sridharan, Compound TCP: a scalable and TCP-friendly congestion control for high-speed networks, *Proceedings of PFLDnet* (2006)

Chapter 3
Joint Congestion Control and Link Scheduling for Internet with Ad-Hoc Wireless Links

In Chap. 2, we considered the Internet with cellular wireless links. A queueing-delay based MaxWeight scheduling (2.14) was proposed for wireless links. This scheduler requires centralized implementation and it is efficient only for cellular wireless networks such as WiMax or WCDMA HSDPA systems. As shown in Fig. 3.1, there is a growing need for Internet access through ad-hoc wireless links enabled by e.g., IEEE 802.11 WiFi, 802.15.4 ZigBee, or 1451 sensor networks, where carrier sense multiple access (CSMA) is employed to allow random access of mobile users for greater flexibility and freedom. For multi-hop links, finding the MaxWeight scheduling policy is in general NP-complete, even with centralized implementation [1]. Some low-complexity algorithms are proposed to approximate the MaxWeight scheduling for ad-hoc networks [2–4]. For these suboptimal algorithms, distributed operation is possible only with the help of message passing. In other words, the MaxWeight scheduler (2.14) is not in the design space of the ad-hoc wireless networks, which prefer distributed scheduling policy without explicit message passing.

Recently, a novel distributed CSMA algorithm was proposed for scheduling of ad-hoc wireless links in [5–7]. Capitalizing on the carrier sensing, it was shown that this asynchronous random access algorithm could achieve maximal throughput without message passing, when all packets traverse only one link (i.e., single-hop) before they leave the network. A joint design of flow congestion control and CSMA scheduling was also developed under the NUM framework for multi-hop wireless networks in [5], where *queue-length* exchanges are *required* among nodes. By integrating this elegant CSMA approach into the proposed window-based cross-layer optimization framework, we next develop TCP and CSMA optimization *without message passing* for the Internet with ad-hoc wireless links.

X. Wang, *Scheduling and Congestion Control for Wireless Internet*,   23
SpringerBriefs in Electrical and Computer Engineering,
DOI: 10.1007/978-1-4614-8420-2_3, © The Author(s) 2014

Fig. 3.1 Internet with a wired
backbone and ad-hoc wireless
links

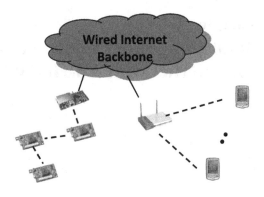

3.1 Network Modeling, Problem Formulation, and Existing Solutions

Consider the Internet with a wired backbone and a CSMA wireless network as in
Fig. 3.1. The wireless network can consist of several separated (IEEE 802.11 WiFi,
802.15.4 ZigBee, or 1451 sensor) sub-networks without loss of generality. The set
of links $L = L_f \cup L_w$ is composed of a wired set L_f and a wireless set L_w. Any
wired link $l \in L_f$ is assumed to have a constant and independent capacity c_l. When
interference is absent, a wireless link $l \in L_w$ has an ergodic capacity b_l. The wireless
links possibly interfere with each other. The link interference can be modeled by a
conflict graph $\mathscr{G}$ with vertices being all wireless links. Two links interfere if, and
only if, there is an edge between them in $\mathscr{G}$; see an example in Fig. 3.2a.

For a given $\mathscr{G}$, we can find a total of $I + 1$ independent sets (ISs) $i = 0, \ldots, I$,
where each IS (not necessarily maximal IS) contains a set of non-interfering links
that are allowed to be simultaneously active. Denote the ith IS by a boolean vector
ξ^i where the lth element of ξ^i, $\xi^i_l = 1$, if link l is in the set, and $\xi^i_l = 0$, otherwise.
As shown in Fig. 3.2b, there always exist a 0th IS, say ξ^0, where $\xi^0_l = 0$, $\forall l \in L_w$
(i.e., no links are active), and the $|L_w|$ ISs, $\xi^i = e^i$, $i = 1, \ldots, |L_w|$, where e^i denotes
a $|L_w|$-dimensional standard basis vector with only the ith element equal to 1 and
all other elements equal to 0 (i.e., only link i is active).

3.1.1 Idealized CSMA Protocol

For completeness, we review an idealized model of CSMA due to [8] and widely
adopted by e.g., [5, 7, 9]. In this model, the carrier-sensing time is assumed negligible. If the transmitter of a link l senses the transmissions of any other links,
it remains silent. When none of its interfering links is sensed active, the transmitter then independently backs off (i.e., waits) for a random period of time that is
continuously distributed with mean $1/a_l$ before transmitting. Due to the negligible
sensing time and continuous distribution of back-off times, the probability for two

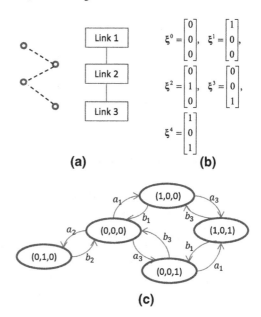

Fig. 3.2 Example: **a** Conflict graph $\mathscr{G}$ for three wireless links. **b** IS vectors ξ^i, $i = 0, 1, \ldots, 4$. **c** CSMA Markov chain

interfering links to transmit at the same time is zero; i.e., the collisions can be avoided. With such an idealized CSMA, the transmission statuses of wireless links form a continuous-time Markov chain, where each state can be described by an IS vector ξ^i (since simultaneous transmission of interfering links never occurs); see Fig. 3.2c.

In the CSMA Markov chain, if link l in a state ξ^i is not active ($\xi_l^i = 0$) and all of its interfering links are not active either, then state ξ^i transits to a new sate $\xi^i + e^l$ with rate a_l; i.e., its lth element changes from 0 to 1, while all other elements remain unchanged. Since link l has an ergodic capacity b_l (i.e., service time $1/b_l$), the state $\xi^i + e^l$ transits back to ξ^i with rate b_l. Define the "transmission aggressiveness" (TA) $\rho_l = \log(a_l/b_l)$. For a given positive vector $\rho := \{\rho_l, \forall l \in L_w\}$, the stationary distribution of the feasible states ξ^i, $i = 0, \ldots, I$, is given by [5, 9]:

$$\Pr(\xi^i; \rho) = \frac{\exp(\sum_{l \in L_w} \rho_l \xi_l^i)}{\sum_j \exp(\sum_{l \in L_w} \rho_l \xi_l^j)}. \tag{3.1}$$

In fact, it can be easily checked from (3.1) that:

$$\frac{\Pr(\xi^i + e^l; \rho)}{\Pr(\xi^i; \rho)} = \exp(\rho_l) = \frac{a_l}{b_l},$$

which is exactly the balance equation between any two feasible states ξ^i and $\xi^i + e^l$. This confirms that (3.1) is invariant, and it is a steady distribution of CSMA Markov chain.

With the stationary distribution (3.1), the probability for link l being active is clearly: $\sum_{i=0}^{I}[\xi_l^i \Pr(\xi^i; \rho)]$; thus the "effective" capacity of link l is:

$$R_l = b_l \sum_{i=0}^{I}[\xi_l^i \Pr(\xi^i; \rho)].$$

3.1.2 Problem Formulation

Again, suppose that the network is shared by a set of unicast flows, identified by their sources $s \in S$. Let $L(s) \subseteq L$ be the set of links that source s uses, and $S(l) \subseteq S$ the set of sources that use link l. Let x_s denote the sending rate of source s. For congestion control, consider a weighted proportionally-fair utility function $p_s \log x_s$ for each flow, as in the TCP Vegas and FAST TCP [10, 11]. For the CSMA wireless network, the scheduling policy is to control the random back-off time of each link in order to attain high "effective link-capacities". In the idealized CSMA model, let $u := \{u_i, i = 0, \ldots, I\}$ denote the stationary distribution of feasible states ξ^i. It was shown that maximization of the (concave) information entropy of u, i.e., $-\sum_{i=0}^{I} u_i \log u_i$, facilitates development of "throughput optimal" distributed CSMA scheduling schemes [5]. For a joint design of congestion control and CSMA scheduling, we then consider the following network optimization problem:

$$\max_{x,\,u} \quad \sum_{s \in S} p_s \log x_s - \sum_{i=0}^{I} u_i \log u_i$$

$$\text{s. t.} \quad \sum_{s \in S(l)} x_s \leq c_l, \ \forall l \in L_f$$

$$\sum_{s \in S(l)} x_s \leq b_l \sum_{i=0}^{I}(u_i \xi_l^i), \ \forall l \in L_w \tag{3.2}$$

$$x_s \geq 0, \ \forall s, \quad 0 \leq u_i \leq 1, \ \forall i, \quad \sum_{i=0}^{I} u_i = 1.$$

Compared to classic NUM formulation, the objective contains the entropy of u in addition to aggregate utility of source rates. Since the entropy is bounded by $\log(I + 1)$, the solution of (3.2) has at most a constant optimality gap $\log(I + 1)$ with classic NUM solutions [5, 7]. This gap can be made (arbitrarily) relatively small when (very) large weights p_s are used for the utility functions.

The formulation (3.2) is different from that in [5] in several senses. While [5] considered a purely wireless network, a hybrid wired and wireless network is considered here, which is more suited for wireless Internet applications. When interference is absent, all links of the CSMA network were assumed to have unit capacity $b_l = 1$

in [5], whereas b_l can be different in (3.2), in order to reflect the rate adaption at the physical layer in practical CSMA networks. More importantly, [5] adopted a node-based formulation as with [1, 12], whereas (3.2) is a link-based one as in [11, 13, 14]. Node-based formulation could facilitate a celebrated "queue-backpressure" type dynamic multi-path routing algorithm, at the expense of requiring each link transmitter to maintain separate queues per flow. On the other hand, the link-based one (3.2) only assumes that each transmitter has a single queue for all incoming traffic; it is more efficient from the implementation viewpoint, for the Internet is currently not supporting multi-path routing. In addition, as will be shown in the sequel, such a link-based formulation (3.2) can also help reduce and/or remove message passing.

3.1.3 Queue-Length Based Solutions

With a node-based formulation, queue-length based joint congestion control and scheduling scheme for CSMA wireless networks was developed in [5]. Relying on "queue-backpressure" based scheduling, the resultant scheme needs to maintain separate queues for each flow at each node, and requires frequent message passing of queue lengths among neighboring nodes. In order to partly circumvent these limitations, we now derive a similar queue-length based solution for the link-based formulation (3.2). It is easy to see that (3.2) is a convex optimization. Let $\lambda := \{\lambda_l, \forall l\}$ denote the Lagrange multipliers for the link capacity constraints. The Lagrangian function of (3.2) is:

$$
\begin{aligned}
L(\lambda; x, u) \\
= \sum_s p_s \log x_s - \sum_{l \in L_f} \lambda_l \left(\sum_{s \in S(l)} x_s - c_l \right) \\
- \sum_i u_i \log u_i - \sum_{l \in L_w} \lambda_l \left(\sum_{s \in S(l)} x_s - b_l \sum_i (u_i \xi_l^i) \right) \\
= \sum_{l \in L_f} \lambda_l c_l + \sum_s (p_s \log x_s - \lambda^s x_s) + \sum_i \left[-u_i \log u_i + u_i \sum_{l \in L_w} \left(\lambda_l b_l \xi_l^i \right) \right]
\end{aligned}
$$

where we define $\lambda^s := \sum_{l \in L(s)} \lambda_l$.
   The dual problem for (3.2) is then:

$$
\min_{\lambda \geq 0} D(\lambda), \quad \text{where} \quad D(\lambda) := \max_{x, u} L(\lambda; x, u). \tag{3.3}
$$

Since (3.2) is a convex optimization, its optimal solution $\{x^*, u^*\}$ and the optimal λ^* for its dual problem (3.3) satisfy the KKT conditions [15]:

$$\frac{p_s}{x_s^*} = \lambda^{s*} := \sum_{l \in L(s)} \lambda_l^*, \quad \forall s \in S \tag{3.4}$$

$$\lambda_l^* \left(\sum_{s \in S(l)} x_s^* - c_l \right) = 0, \quad \sum_{s \in S(l)} x_s^* \leq c_l, \quad \forall l \in L_f \tag{3.5}$$

$$\lambda_l^* \left(\sum_{s \in S(l)} x_s^* - b_l \sum_i [u_i^* \xi_l^i] \right) = 0, \quad \sum_{s \in S(l)} x_s^* \leq b_l \sum_i [u_i^* \xi_l^i], \quad \forall l \in L_w \tag{3.6}$$

$$u_i^* = \frac{\exp(\sum_{l \in L_w} (\lambda_l^* b_l \xi_l^i))}{\sum_j \exp(\sum_{l \in L_w} (\lambda_l^* b_l \xi_l^j))}, \quad \forall i. \tag{3.7}$$

The optimal $\{x^*, u^*, \lambda^*\}$ satisfying (3.4–3.7), can be obtained by classic subgradient based dual iterations. For a given λ, the dual-optimal source rate is clearly $x_s^*(\lambda) = p_s/\lambda^s$, $\forall s$. It can be also shown from the optimality condition that, $\forall i$,

$$u_i^*(\lambda) = \frac{\exp(\sum_{l \in L_w} (\lambda_l b_l \xi_l^i))}{\sum_j \exp(\sum_{l \in L_w} (\lambda_l b_l \xi_l^j))}. \tag{3.8}$$

Given $x_s^*(\lambda)$ and $u_i^*(\lambda)$, the dual problem of (3.2) can be solved by the sub-gradient descent iterations:

$$\lambda_l[t+1] = \left[\lambda_l[t] + \varepsilon \left(\sum_{s \in S(l)} x_s^*(\lambda[t]) - c_l \right) \right]^+, \quad \forall l \in L_f,$$

$$\lambda_l[t+1] = \left[\lambda_l[t] + \varepsilon \left(\sum_{s \in S(l)} x_s^*(\lambda[t]) - b_l \sum_i [u_i^*(\lambda[t]) \xi_l^i] \right) \right]^+, \quad \forall l \in L_w \tag{3.9}$$

where t is the iteration index, ε is a stepsize, and $[x]^+ := \max(0, x)$. Convergence of (3.9) to the optimal λ^* is guaranteed from any initial $\lambda[0] \geq 0$ [15]. Due to the zero duality gap, the corresponding $x_s^*(\lambda^*)$ and $u_i^*(\lambda^*)$ yield the optimal solutions for the primal problem (3.2). Based on this sub-gradient method, a joint congestion control and CSMA scheduling can be devised as follows.

Comparing (3.8) with (3.1), we find that the distribution $u_i^*(\lambda)$ can be actually obtained as the steady state distribution of the idealized CSMA network with the TA of each link set to $\rho_l \equiv \lambda_l b_l$. On the other hand, with the aggregate arrival rate $\sum_{s \in S(l)} x_s^*(\lambda[t])$ and service rate c_l or $b_l \sum_i [u_i^*(\lambda[t]) \xi_l^i]$ for each link, the subgradient iterate in (3.9) can be seen as an evolution of the scaled queue lengths εQ_l:

$$\lambda_l[t+1] \equiv \varepsilon Q_l[t+1]$$

$$= \begin{cases} \left[\varepsilon\left[Q_l[t] + \left(\sum_{s \in S(l)} x_s^*(\varepsilon Q[t]) - c_l\right)\right]^+, & \forall l \in L_f, \\ \left[\varepsilon\left[Q_l[t] + \left(\sum_{s \in S(l)} x_s^*(\varepsilon Q[t]) - b_l \sum_i [u_i^*(\varepsilon Q[t])\xi_l^i]\right)\right]^+, & \forall l \in L_w. \end{cases}$$

As a result, we propose the following congestion control and CSMA scheduling schemes:

- *Congestion control*: In the update-period t per flow s, the source sets its rate $x_s[t] = \frac{p_s}{\varepsilon \sum_{l \in S(l)} Q_l[t]}$.
- *CSMA scheduling*: In the update-period t per wireless link, the link-transmitter sets its TA as $\rho_l[t] = \varepsilon b_l Q_l[t]$, or equivalently, sets the mean of its random back-off time as $e^{-\varepsilon b_l Q_l[t]}/b_l$, such that the effective link capacity becomes $R_l = b_l \sum_i [u_i^*(\varepsilon Q[t])\xi_l^i]$.

These schemes together with the queue evolution $Q_l[t]$, then follow the sub-gradient approach (3.9) to solve (3.2).

   In the above joint design, the CSMA scheduling is implemented in a truly distributed manner as each wireless link-transmitter uses local queue length Q_l to set its random back-off timer without centralized coordination and any message passing from neighbors. However, the congestion controller at the source node would require message passing of the (aggregate) queue lengths $\sum_{l \in S(l)} Q_l$ along the flow route from intermediate nodes. Compare to the proposed scheme in [5], where a "queue backpressure" mechanism is used to indirectly transfer the congestion prices at the intermediate nodes along the flow route to the source in a hop-by-hop manner. As a result, the sources can adapt their rates based on local congestion prices (i.e. queue lengths) without the need of message passing. However, in the resultant CSMA scheduling, each wireless node uses the queue differences between itself and all its one-hop neighbors to set its random back-off timer; this requires the message passing of queue lengths among all neighboring nodes. In short, for both joint congestion control and CSMA scheduling schemes, one-hop message passing of queue lengths of all active links is required. Recall that in order to perform the queue-backpressure schemes, each node needs to maintain separate queues for each flow through it. Suppose that each active link is used by $N \geq 1$ flows on average. Then compared to the proposed scheme for (3.2), the schemes stemming from node-based formulation in [5] need to maintain N times more queues and incur N times more message passing overheads.

   Compared to its counterpart in [5], the proposed queue-length based congestion control and CSMA solution for link-based formulation (3.2) has independent value and certain novelty. Yet, we still regard it as an "existing" solution since it can be readily derived via applying the approach in [5] under classic NUM framework. The proposed congestion control is implemented by a source-rate controller (instead of TCP window-control), and explicit message passing from the network is required; hence, they are difficult to operate over current Internet infrastructure. To overcome these limitations, we next show that jointly optimal, yet readily deployable, TCP congestion control and CSMA scheduling can be actually developed for (3.2), when queueing delays (instead of queue lengths) are used to play the role of Lagrange multipliers.

3.2 Joint Congestion Control and CSMA Scheduling Without Message Passing

To guide our design, we consider an idealized CSMA protocol and a fluid model of network where packets are infinitely divisible and small [16].

In this idealized network fluid model, let $w := \{w_s, \forall s\}$ collect the window sizes for all sources $s \in S$, $q := \{q_l, \forall l\}$ collect the round-trip queueing delays for all links $l \in L$, and $d := \{d_s, \forall s\}$ collect the fixed round-trip propagation (plus processing) delays for all sources. For the CSMA wireless network, let $R := \{R_l, \forall l \in L_w\}$ denote the effective capacity vector for wireless links under the given scheduling strategy. Upon defining the aggregate queueing delays $q^s := \sum_{l \in L(s)} q_l$ along the flow routes, we have the following relationships for source rates x_s, window sizes w_s, and queueing delays q_l [cf. (2.6–2.8)]:

$$x_s(d_s + q^s) = w_s, \quad \forall s \in S \tag{3.10}$$

$$q_l\left(\sum_{s \in S(l)} x_s - c_l\right) = 0, \quad \sum_{s \in S(l)} x_s \le c_l, \quad \forall l \in L_f \tag{3.11}$$

$$q_l\left(\sum_{s \in S(l)} x_s - R_l\right) = 0, \quad \sum_{s \in S(l)} x_s \le R_l, \quad \forall l \in L_w. \tag{3.12}$$

3.2.1 Joint Design in Network Fluid Model

We next propose a joint TCP congestion control and CSMA scheduling design in the (simplified) fluid model. Comparing the fluid model identities (3.11, 3.12) with the KKT conditions (3.5, 3.6), it follows that the queueing delay q_l in the fluid model can play a similar role of Lagrange multiplier λ_l associated with the link capacity constraints in (3.2). Then implied by (3.7), we propose a CSMA scheduling strategy, where each link actually employs the queueing delay to set its TA as $\rho_l = b_l q_l$ such that the resultant steady state distribution is [cf. (3.1) and (3.8)]:

$$\Pr(\xi^i) = \frac{\exp(\sum_{l \in L_w} b_l q_l \xi_l^i)}{\sum_j \exp(\sum_{l \in L_w} b_l q_l \xi_l^j)} = u_i^*(q), \tag{3.13}$$

and the effective link capacities are:

$$R_l = b_l \sum_i [u_i^*(q)\xi_l^i], \quad \forall l \in L_w. \tag{3.14}$$

On the other hand, in the TCP congestion control, each flow source adjusts its transmission window size to prevent network congestion according to a locally observable congestion measure. Using delay as the congestion measure, we consider adopting the QUIC-TCP algorithm in Chap. 2 to adjust the window size per flow s. Denote the (local) total round-trip delay of flow s as $\bar{d}_s := d_s + q^s$. With $x_s = w_s/\bar{d}_s$ from (3.10), define $v_s := w_s - x_s d_s - p_s$. Parameterized by a constant $\rho \in [0, 1]$, QUIC-TCP entails a class of end-to-end algorithms with window adjustment following the ODEs:

$$\frac{d}{dt}w_s(t) = -\frac{d_s}{\bar{d}_s}w_s^{-2\rho+1}v_s, \quad \forall s. \tag{3.15}$$

The proposed joint design adjusts the TCP window size w_s and the CSMA TA ρ_l, in order to control the source-rate x, wireless-link capacity R and queueing delay q. The goal of the proposed network control is to drive the network operating point towards a desired equilibrium that yields the optimal solution for (3.2). Define $v := \{v_s, \forall s\}$. In equilibrium of window update (3.15), we clearly have $v_s = 0, \forall s$; i.e., $v = 0$. Let w^* denote the window-size vector with this equilibrium, and x^*, u^*, and q^* the corresponding source-rate, CSMA stationary distribution, and queueing-delay vectors. We can establish the following theorem:

Theorem 3.1 *For the proposed joint congestion control and CSMA scheduling scheme, there is a unique window size vector w^* such that $v = 0$, and the corresponding rate vector x^* and CSMA steady distribution u^* are the optimal ones for (3.2).*

Proof See Appendix 3A.

Theorem 3.1 states that the unique equilibrium of the proposed joint window adjustment and CSMA scheduling scheme leads to the optimal x^* and u^* for (3.2). Based on local observations without message passing, the proposed scheme aims to entail an *implicit primal-dual* update of $\{x, u, q\}$ towards the optimal equilibrium that solves (3.2). Different from the prior queue-length based NUM solutions, the proposed scheme fits well into the design space of the Internet protocols and can be thus readily deployed with today's Internet infrastructure. Also different from the heuristic schemes, the global convergence of the proposed scheme to its unique equilibrium can be analytically established using the Lyapunov method.

3.2.2 Convergence Analysis

For convenience, rewrite the problem (3.2) as:

$$\max_{x,\,u} \quad \sum_s p_s \log x_s - \sum_i u_i \log u_i$$

$$\text{s. t.} \quad A_f x \le c, \quad A_w x \le R, \tag{3.16}$$

$$x_s \ge 0, \ \forall s, \quad 0 \le u_i \le 1, \ \forall i, \quad \sum_i u_i = 1.$$

where $c := \{c_l, \forall l \in L_f\}$, $R := \{R_l, \forall l \in L_w\}$ with $R_l = b_l \sum_i (u_i \xi_l^i)$, and the routing matrix $A := [A_f^T, \ A_w^T]^T$ with its (l, s)th entry $A_{ls} = 1$, if $s \in S(l)$ and $A_{ls} = 0$, otherwise.

Consider the idealized network fluid model. In the proposed CSMA scheduling scheme, the queueing delays q_l of wireless links are employed for random access to result the link capacities R_l (3.14). Let $q_w := \{q_l, \forall l \in L_w\}$, and $J_{R|q_w} := \{\frac{\partial R_l}{\partial q_n}, \forall l, n\}$ denotes the Jacobian matrix of vector R with respect to vector q_w. Then we can establish the following lemma:

Lemma 3.1 *With the proposed CSMA scheduling strategy, the Jacobian matrix $J_{R|q_w}$ is positive definite for any queueing delay vector $q_w > 0$.*

Proof See Appendix 3B.

Lemma 3.1 shows that using queueing delays to set the TA of links in idealized CSMA, the wireless links are coupled in a "desired" manner such that $J_{R|q_w}$ is positive definite. Using this nice property, we can then show the following global stability result in the idealized network fluid model:

Theorem 3.2 *The unique equilibrium $v = 0$ of the proposed joint QUIC-TCP congestion control and CSMA scheduling scheme is globally asymptotically stable for $0 \le \rho \le 1$; i.e., the proposed scheme globally converges to its unique equilibrium that yields the optimal solution for (3.2).*

Proof See Appendix 3C.

Theorem 3.2 establishes the global stability of the equilibrium $v = 0$; as a result, the TCP window sizes in (3.15) globally converge to its unique equilibrium w^*, and, consequently, the source rates and CSMA steady distribution converge to the optimal x^* and u^* for (3.2). This theorem proves the existence of jointly optimal TCP congestion control and CSMA scheduling schemes for Internet applications over ad-hoc wireless links, *without explicit message passing*. The key is that we follow a novel cross-layer optimization approach to develop the proposed schemes as decomposition parts of the implicit primal-dual solvers for relevant network optimization, with queueing delays playing the roles of Lagrange multipliers.

3.2.3 Development of Practical Schemes

Using the insights provided by the joint design in the network fluid model, we next develop the congestion control and CSMA scheduling schemes for practical TCP and IEEE 802.11 protocol.

Congestion control: Adopt the QUIC-TCP implementation in Chap. 2. Whenever an in-order ACK is received, each flow source s computes the RTT, RTT_s, of the acknowledged packet, Using RTT_s, the source then updates the values of $BaseRTT_s$ and $AvgRTT_s$. The window control component then adjusts the transmission window size w_s as [cf. (3.15)]:

$$w_s \leftarrow w_s - \kappa \frac{BaseRTT_s}{AvgRTT_s} w_s^{-2\rho+1} \left(w_s - \frac{BaseRTT_s}{AvgRTT_s} w_s - p_s \right). \qquad (3.17)$$

When an in-order ACK is not received, the slow start and/or fast recovery schemes in standard TCP are preserved for time-outs or duplicate ACKs.

CSMA scheduling: In the proposed CSMA scheduling strategy (at link-layer), each wireless link transmitter employs the queueing delay to set its TA. To estimate its queueing delay, let each wireless transmitter node read its queue length $QueLEN_l[t]$ and calculate the average rate $AveR_l[t]$ using a low-pass filter at an update time t. Implied by the Little's law, the average queueing delay is given by the ratio of average queue length and average rate. Based on the fluid limit argument [17], the "stochastic" delay $\frac{QueLEN_l[t]}{AveR_l[t]}$ could play the role of average delay q_l in the deterministic fluid model. Supposing that the interference-free ergodic capacity b_l is known (e.g., informed by a bandwidth estimator at the physical layer), we then propose that the transmitter of link l sets its TA as:

$$\rho_l[t] = b_l \frac{QueLEN_l[t]}{AveR_l[t]}, \qquad (3.18)$$

or equivalently, it sets the mean of its backoff time as $\exp(-b_l \frac{QueLEN_l[t]}{AveR_l[t]})/b_l$, to randomly access the channel. In the IEEE 802.11 CSMA protocol, after a successful transmission without collisions, the wireless transmitter resets its contention window size to a predetermined value CW_{min} and randomly selects an integer c from $\{0, 1, \ldots, CW_{min} - 1\}$. If the channel is sensed idle, the transmitter then backs off a time of cT_{slot} before next transmission, where T_{slot} is a predefined minslot duration. In this case, the mean backoff time is $\frac{CW_{min}-1}{2} T_{slot}$. To incorporate the proposed strategy, we let the transmitter of link l set its own contention window size to $2\exp(-b_l \frac{QueLEN_l[t]}{AveR_l[t]})/(b_l T_{slot}) + 1$ (instead of CW_{min}) after a successful transmission, and perform aforementioned random backoff such that the mean backoff time becomes $\exp(-b_l \frac{QueLEN_l[t]}{AveR_l[t]})/b_l$. Except for this modification, all other components of IEEE 802.11 CSMA are preserved.

Clearly, the proposed congestion control and scheduling schemes are confined to the design space of TCP and CSMA protocols. While the congestion controller (3.17) preserves the distributed end-to-end window-control mechanism of TCP, the CSMA random back-off strategy (3.18) can be performed in a truly distributed manner, without message passing. They can be implemented in asynchronous manners and can be readily deployed with the current (layered) Internet infrastructure. In addition, as they are designed based on the fluid-model counterparts that are provably optimal for (3.2), high performance can be expected for these schemes.

3.3 Summary

We develop joint TCP congestion control and CSMA scheduling schemes for the Internet traffic over distributed multi-hop wireless links without the need of explicit message passing. To guide the development of the desired window-based primal-dual solvers, we perform an analysis based on the ideal CSMA modeling at the design time. With queueing delays playing the role of Lagrange multipliers, we anticipate from the optimality conditions of the problem that $Y(w) = (1/2) \sum_{s \in S} (v_s/(w_s)^\rho)^2$ could still serve as a Lyapunov function in the network fluid model. By checking the first time-derivative $\frac{d}{dt} Y(w(t))$, it is found that the QUIC-TCP can be still adopted for flow congestion control. To guarantee the asymptotic stability of the QUIC-TCP ODEs (3.15), it follows from the insights in Chap. 2 that the effective wireless-link capacity under the proposed CSMA scheduling strategy should be made coupled in a "desired" manner; i.e., the Jacobian matrix of effective wireless-link capacities with respect to queueing delays, $J_{R|q_w}$, needs to be positive definite such that the QUIC-TCP congestion control and CSMA scheduling could entail an implicit primal-dual solver to the intended optimization problem. To obtain such a CSMA scheme, we borrow the ideas of distributed CSMA in [5]. Since queueing delays act as Lagrange multipliers in our design, we naturally propose an CSMA algorithm where each link-transmitter employs its local queueing delay to properly control its mean back-off time to randomly access the shared wireless channel in a truly *distributed* manner. Relying on the Lyapunov argument, we rigorously prove the global convergence/stability of proposed joint QUIC-TCP and CSMA scheduling over ad-hoc wireless links. The proposed algorithms are readily deployed for practical TCP and CSMA schemes in IEEE 802.11, 802.15.4 standards.

Appendix 3A: Proof of Theorem 3.1

The proof mimics that for Theorem 2.1. At the equilibrium $v = 0$ with the proposed QUIC-TCP, we have $w_s^* - x_s^* d_s - p_s = 0$; on the other hand, it follows from the fluid model identity (3.10) that $w_s^* = x_s^*(d_s + q^{s*})$. Hence, we readily have $p_s = x_s^* q^{s*}$, $\forall s$. This is exactly the KKT condition (3.4), if we let the optimal dual vector $\lambda^* \equiv q^*$. With this equivalence mapping, the fluid model identities (3.11, 3.12) together with (3.13, 3.14) under the proposed CSMA scheduling, also become the KKT conditions (3.5–3.7). Since x^*, u^* and q^* satisfy the sufficient and necessary KKT optimality conditions, they are the optimal solutions of (3.2) and its dual problem (3.3). Furthermore, we can show the uniqueness of mapping from w^* to x^* under (3.11–3.14). Due to this uniqueness, as well as the existence and uniqueness of x^* for (3.2), it follows that the optimal window equilibrium w^* always exists and it is unique.

Appendix 3B: Proof of Lemma 3.1

From (3.13), we find the partial derivative: $\forall n \in L_w$,

$$\frac{\partial u_i(q)}{\partial q_n} = \frac{b_n \xi_n^i \exp(\sum_l [q_l b_l \xi_l^i])}{\sum_j \exp(\sum_l [q_l b_l \xi_l^j])} - \frac{\exp(\sum_l [q_l b_l \xi_l^i]) \sum_j [b_n \xi_n^j \exp(\sum_l q_l b_l \xi_l^j)]}{[\sum_j \exp(\sum_l [q_l b_l \xi_l^j])]^2}$$

$$= b_n u_i(q) \xi_n^i - u_i(q) \left(\frac{b_n \sum_j [\xi_n^j \exp(\sum_l [q_l b_l \xi_l^j])]}{\sum_{j'} \exp(\sum_l [q_l b_l \xi_l^{j'}])} \right)$$

$$= u_i(q)(b_n \xi_n^i - R_n).$$

It in turn follows from (3.14) that: $\forall l, \; n \in L_w$,

$$\frac{\partial R_l}{\partial q_n} = b_l \sum_i \left(\frac{\partial u_i(q)}{q_n} \xi_l^i \right)$$

$$= b_l \sum_i [u_i(q)(b_n \xi_n^i - R_n) \xi_l^i]$$

$$= b_l b_n \sum_i [u_i(q) \xi_l^i \xi_n^i] - b_l \left(\sum_i [u_i(q) \xi_l^i] \right) R_n$$

$$= b_l b_n \left[\sum_i [u_i(q) \xi_l^i \xi_n^i] - \left(\sum_i [u_i(q) \xi_l^i] \right) \left(\sum_i [u_i(q) \xi_n^i] \right) \right]. \qquad (3.19)$$

Suppose that there are $I + 1$ ISs for the given conflict graph $\mathcal{G}$ of the wireless links. Recall that there is always a 0th IS, $\xi^0 = [0 \, 0 \, \ldots \, 0]^T$. This implies that

$$\sum_{i=0}^{I} [u_i(q) \xi_l^i \xi_n^i] = \sum_{i=1}^{I} [u_i(q) \xi_l^i \xi_n^i],$$

$$\sum_{i=0}^{I} [u_i(q) \xi_l^i] = \sum_{i=1}^{I} [u_i(q) \xi_l^i];$$

i.e., we can omit $i = 0$ for the summations in (3.19).

Define a vector $\mu := [u_1(q) \ \ldots \ u_I(q)]^T$, a diagonal matrix $U := \text{diag}(\mu)$, and let matrix $\Upsilon := [\xi^1 \ \ldots \ \xi^I]^T$ collect the IS vectors for the given conflict graph of wireless links. Note that $u_0(q)$ and ξ^0 are not included in the vector and matrices. Furthermore, let $b := \{b_l, \forall l \in L_w\}$ and $B := \text{diag}(b)$. Using μ, Υ, and B, it then follows from (3.19) that the Jacobian matrix $J_{R|q_w} = \{\frac{\partial R_l}{\partial q_n}, \forall l, n\}$ is given by:

$$J_{R|q_w} = B^T \Upsilon^T U \Upsilon B - B^T \Upsilon^T \mu \mu^T \Upsilon B = B^T \Upsilon^T (U - \mu \mu^T) \Upsilon B.$$

By the definitions of μ and U, we have:

$$U - \mu \mu^T = \begin{bmatrix} u_1(1-u_1) & -u_1 u_2 & \cdots & -u_1 u_I \\ -u_2 u_1 & u_2(1-u_2) & \cdots & -u_2 u_I \\ \vdots & \vdots & \ddots & \vdots \\ -u_I u_1 & -u_I u_2 & \cdots & u_I(1-u_I) \end{bmatrix}$$

Now check the diagonal dominance of matrix $U - \mu \mu^T$: for each row i,

$$|u_i(1-u_i)| - \sum_{j \neq i, j=1}^{I} |-u_i u_j| = u_i(1-u_i) - \sum_{j \neq i, j=1}^{I} u_i u_j$$

$$= u_i \left(1 - \sum_{j=1}^{I} u_j\right) = u_1 u_0 > 0, \qquad (3.20)$$

since $u_0 = \dfrac{1}{\sum_j e^{\sum_l [q_l b_l \xi_l^j]}} > 0$ and $u_i = \dfrac{e^{\sum_l [q_l b_l \xi_l^j]}}{\sum_j e^{\sum_l [q_l b_l \xi_l^j]}} > 0$.

Since the symmetric $U - \mu \mu^T$ is strictly diagonally dominant with positive diagonal entries, it is positive definite. Recall that the $I \times |L_w|$ matrix Υ must have full column rank, i.e., $\text{rank}(\Upsilon) = |L_w|$, since it always contains the $|L_w|$ standard basis vectors $\xi^i = e^i, i = 1, \ldots, |L_w|$. Because B is a $|L_w| \times |L_w|$ diagonal matrix with positive diagonals, the matrix ΥB has full column rank. This implies that for any non-zero vector v, $\Upsilon B v$ is a non-zero vector. Then for any $v \neq 0$,

$$v^T J_{R|q_w} v = v^T [B^T \Upsilon^T (U - \mu \mu^T) \Upsilon B] v = (\Upsilon B v)^T (U - \mu \mu^T)(\Upsilon B v) > 0,$$

due to the positive-definiteness of $U - \mu \mu^T$. The lemma readily follows.

Appendix 3C: Proof of Theorem 3.2

For a given w, let $\mathscr{B}$ denote the set of bottleneck links. Define A_B, q_B, c_B, and R_B the corresponding sub-matrix or sub-vectors of q, c, and R for bottleneck links. Let the diagonal matrices $X := \text{diag}(x)$, $W := \text{diag}(w)$, $D := \text{diag}(d)$, and $\bar{D} := \text{diag}(\bar{d})$ where $\bar{d} := \{\bar{d}_s, \forall s\}$. We can rewrite (3.10) in the matrix form:

$$X(A_B^T q_B + d) = w. \tag{3.21}$$

Differentiating both sides of (3.21) with respect to w and then multiplying both sides by $A_B \bar{D}^{-1}$ yield:

$$A_B J_{x|w} + A_B \bar{D}^{-1} X A_B^T J_{q_B|w} = A_B \bar{D}^{-1}. \tag{3.22}$$

Partition the bottleneck-only routing matrix A_B and queueing delay vector q_B into two parts:

$$A_B = \begin{bmatrix} A_B^f \\ A_B^w \end{bmatrix}, \qquad q_B = \begin{bmatrix} q_B^f \\ q_B^w \end{bmatrix}$$

For the wired bottleneck links, it holds $A_B^f x = c_B$; thus $A_B^f J_{x|w} = 0$. For wireless bottleneck links, $A_B^w x = R_B$ implies:

$$A_B^w J_{x|w} = J_{R_B|w} = [0 \;\; J_{R_B|q_B^w}] J_{q_B|w}.$$

Then overall we have:

$$A_B J_{x|w} = \begin{bmatrix} A_B^f \\ A_B^w \end{bmatrix} J_{x|w} = \begin{bmatrix} 0 & 0 \\ 0 & J_{R_B|q_B^w} \end{bmatrix} J_{q_B|w} := N J_{q_B|w}$$

It then follows that:

$$(N + A_B \bar{D}^{-1} X A_B^T) J_{q_B|w} = A_B \bar{D}^{-1}.$$

Since $J_{R_B|q_B^w}$ is positive definite from Lemma 3.1 (or it is 0 if there are no wireless bottlenecks), the matrix N is clearly positive semi-definite. Using the convenient notation: $M := A_B^T (N + A_B \bar{D}^{-1} X A_B^T)^{-1} A_B$, it is clear that the matrix M is positive semi-definite and $J_{x|w} = \bar{D}^{-1}(I - XM\bar{D}^{-1})$.

With $Y(w) = \frac{1}{2} \sum_{s \in S} \left(\frac{v_s}{(w_s)^\rho} \right)^2$, we can then follow the similar lines in the proof of Theorem 2.2 to show $dY(w(t))/dt < 0$, i.e., $Y(w(t))$ is strictly decreasing in t, at all points unless $v = 0$. The unique equilibrium $v = 0$ is thus globally asymptotically stable.

References

1. L. Tassiulas, A. Ephremides, Stability properties of constrained queueing systems and scheduling policies for maximum throughput in multihop radio networks. IEEE Trans. Autom. Control **36**(12), 1936–1948 (1992)

2. C. Joo, X. Lin, N. Shroff, Understanding the capacity region of the greedy maximal scheduling algorithm in multi-hop wireless networks, *Proceedings of IEEE INFOCOM Conference*, pp. 1103–1111, Phoenix, AZ, Apr. 2008
3. M. Leconte, J. Ni, R. Srikant, Improved bounds on the throughput efficiency of greedy maximal scheudling in wireless networks, *Proceedings of ACM MobiHoc Conference*, pp. 165–174, May 2009
4. X. Wu, R. Srikant, Scheduling efficiency of distributed greedy scheduling algorithms in wireless networks, *Proceedings of IEEE INFOCOM Conference*, pp. 1–12, Barcelona, Spain, Apr. 2006
5. L. Jiang, J. Walrand, A distributed CSMA algorithm for throughput and utility maximization in wireless networks. IEEE/ACM Trans. Netw. **18**(3), 960–972 (2010)
6. L. Jiang, D. Shah, J. Shin, J. Walrand, Distributed random access algorithm: scheduling and congestion control. IEEE Trans. Inf. Theory **56**(12), 6182–6207 (2010)
7. J. Liu, Y. Yi, A. Proutiere, M. Chiang, H. Poor, Towards utility-optimal random access without message passing, *Wireless Communications and Mobile Computing* (2009). doi:10.1002/wcm.000
8. F. Kelly, Stochastic models of computer communication systems. J. R. Stat. Soc. **47**(3), 379–395 (1985)
9. J. Wang, D. Wei, S. Low, Modeling and stability of FAST TCP, *Proceedings of IEEE INFOCOM Conference*, pp. 938–948, Miami, FL, 13–17 Mar. 2005
10. L. Brakmo, L. Peterson, TCP Vegas: end-to-end congestion avoidance on a global internet. IEEE J. Sel. Areas Commun. **13**(8), 1465–1480 (1995)
11. D. Wei, C. Jin, S. Low, S. Hedge, FAST TCP: motivation, architecture, algorithms, performance. IEEE/ACM Trans. Netw. **14**(6), 1246–1259 (2006)
12. M. Neely, E. Modiano, C. Li, Fairness and optimal stochastic control for heterogenous networks. IEEE/ACM Trans. Netw. **16**(2), 396–409 (2008)
13. X. Lin, N. Shroff, The impact of imperfect scheduling on cross-layer rate control in wireless networks. IEEE/ACM Trans. Netw. **14**(2), 302–315 (2006)
14. Y. Yu, G. Giannakis, Joint congestion control and OFDMA scheduling for hybrid wireline-wireless networks, *Proceedings of IEEE INFOCOM Conference*, pp. 973–981, Anchorage, AK, 6–12 May 2007
15. S. Boyd, L. Vandenberghe, *Convex Optimization* (Cambridge University Press, Cambridge, 2004)
16. J. Mo, J. Walrand, Fair end-to-end window-based congestion control. IEEE/ACM Trans. Netw. **8**(5), 556–567 (2000)
17. A. Stolyar, Maximizing queueing network utility subject to stability: greedy primal-dual algorithm. Queueing Syst. **50**, 401–457 (2005)

Chapter 4
Generalizations and Interesting Directions

The works in Chaps. 2 and 3 can clearly serve as a stepping stone towards a theoretic foundation for the proposed design-space oriented cross-layer optimization, and practical designs of high-performance network protocols and schemes. We next outline some possible generalizations and interesting directions.

4.1 Internet with Both Cellular and Ad-Hoc Wireless Links

To meet the demand for ubiquitous data services, current mobile users can in fact conveniently access the Internet both through cellular systems with a centralized access point, such as IEEE 802.16 WiMax and UMTS HSDPA networks, and through ad-hoc wireless links enabled by e.g., CSMA-based IEEE 802.11 WiFi networks. Chaps. 2 and 3 addressed the joint design of congestion control and link scheduling for TCP flows over the Internet with cellular or ad-hoc wireless links, respectively. Since the cellular and ad-hoc wireless systems typically operate at different frequency bands in practice, only minimum interference exists between any cellular and ad-hoc wireless links. This implies that the scheduling strategies for cellular and ad-hoc links are decoupled. It then follows that the design and analysis of the proposed schemes for the two cases can be combined to provide solutions for the Internet with both cellular and ad-hoc wireless links.

Without loss of generality, assume an Internet with a wired backbone, a cellular system, and a CSMA ad-hoc network; see Fig. 4.1. The set of logical data links $L = L_f \cup L_c \cup L_a$ in the network consists of a wired link set L_f, a centralized wireless link set L_c, and an ad-hoc wireless link set L_a. Assume that there is no interference between any link in L_c and any link in L_a. While any wired link $l \in L_f$ has a constant and independent link capacity c_l, the capacities for cellular and ad-hoc wireless links follow the models in Chaps. 2 and 3. For a joint design of congestion control and link-scheduling, we consider the network optimization problem:

X. Wang, *Scheduling and Congestion Control for Wireless Internet*,
SpringerBriefs in Electrical and Computer Engineering,
DOI: 10.1007/978-1-4614-8420-2_4, © The Author(s) 2014

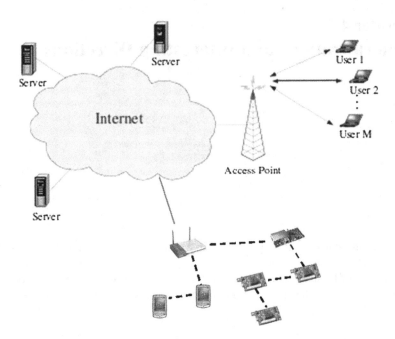

Fig. 4.1 Internet with a wired backbone, cellular and ad-hoc wireless links

$$\max_{x,\,r,\,u}\quad \sum_{s\in S} p_s \log x_s - \sum_{i=0}^{I} u_i \log u_i$$

$$\text{s. t.}\quad \sum_{s\in S(l)} x_s \le c_l,\ \forall l \in L_f$$

$$\sum_{s\in S(l)} x_s \le r_l,\ \forall l \in L_c;\quad r \in \bar{\mathscr{R}} \tag{4.1}$$

$$\sum_{s\in S(l)} x_s \le R_l = b_l \sum_{i=0}^{I} (u_i \xi_l^i),\ \forall l \in L_a$$

$$0 \le u_i \le 1,\ \forall i,\quad \sum_{i=0}^{I} u_i = 1;\quad x_s \ge 0,\ \forall s.$$

Relying on the proposed approach, we can combine the congestion control (at transport-layer of source node) (2.12), centralized link-scheduling (at link-layer of access point) (2.11), and CSMA scheduling (at link-layer of ad-hoc node) strategy (3.13) to solve (4.1) in the network fluid model. As with Theorems 2.1 and 3.1, we have:

Corollary 4.1 *For the proposed joint congestion control and link scheduling scheme, there is a unique window-size vector w^* at its equilibrium, and the corresponding rate vector x^* is the optimal one for (4.1).*

Proof See Appendix 4A.

Combining the proof for Theorems 2.2 and 3.2, we can also establish:

Corollary 4.2 *Under the proposed schedulers, the window-based control (2.12) with $0 < \rho \leq 1$ globally converges to its unique equilibrium w^* and, consequently, the source-rate vector converges to the optimal x^* for (4.1).*

Proof See Appendix 4B.

Inferred by (2.12), (2.11), and (3.13), we can then adopt QUIC-TCP algorithm (2.13) and scheduling schemes (2.14), (3.18) for practical networks. The proposed schemes are readily deployable. In addition, they have performance guarantees since they are designed based on the fluid-model counterparts that are provably optimal for (4.1).

4.2 General Utility Functions

In Chaps. 2 and 3, the proposed schemes are developed for NUM with a proportionally fair utility function. The approach can be generalized to NUM with general concave utility functions. Use the cellular wireless link case for specificity. Consider the NUM problem (2.1) with general concave $U_s(x_s)$. In this case, the KKT conditions (2.3–2.5) preserves, while condition (2.2) becomes:

$$U'_s(x^*_s) = \lambda^{s*} := \sum_{l \in L(s)} \lambda^*_l, \quad \forall s \in S \tag{4.2}$$

where U'_s denotes the first derivative of function U_s. From the fluid model identities (2.6)–(2.9), it follows that the optimal $\{x^*, \bar{r}^*\}$ for (2.1) should satisfy:

$$\begin{cases} \bar{r}^* = \arg\max_{\bar{r} \in \bar{\mathcal{R}}} \sum_{l \in L_w} q^*_l \bar{r}_l, \\ x^*_s(d_s + U'_s(x^*_s)) = w^*_s, \quad \forall s \in S \end{cases} \tag{4.3}$$

where the queueing delay q^*_l plays the role of λ^*_l, and w^*_s denotes the optimal window size.

To attain the optimal network equilibrium, it is clear that the queueing-delay based MaxWeight scheduler (2.11) should be implemented at the wireless access point. On the other hand, redefine $v_s := w_s - x_s d_s - x_s U'_s(x_s)$. As with (2.12), we propose the window-update ODEs:

$$\frac{d}{dt} w_s(t) = -g_s(w_s, d_s, x_s) v_s, \quad \forall s \tag{4.4}$$

where $g_s(w_s, d_s, x_s)$ is a function of local variables w_s, d_s, and x_s. As with Theorem 2.1, it can be shown that (2.11) and (4.4) have a unique equilibrium at $v = 0$ that leads to the optimal solution of (2.1) with general $U_s(x_s)$.

To show the convergence of the proposed joint design to its unique equilibrium, adopt again the quadratic function $Y(w) = (1/2) \sum_{s \in S} (v_s/(w_s)^\rho)^2$ as a candidate Lyapunov function. By checking $\frac{d}{dt} Y(w(t))$, we can show that (U_s'' denotes the second derivative of function U_s):

Proposition 4.1 *A necessary condition for $Y(w)$ to be a Lyapunov function is:*
$g_s(w_s, d_s, x_s) = x_s(d_s + U_s'(x_s) + x_s U_s''(x_s))w_s^{-2\rho}, \forall s.$

Proof See Appendix 4C.

Following Proposition 4.1, we consider the window-update ODEs:

$$\frac{d}{dt} w_s(t) = -x_s(d_s + U_s'(x_s) + x_s U_s''(x_s))w_s^{-2\rho} v_s, \forall s \qquad (4.5)$$

To identify the sufficient conditions that make $Y(w)$ a Lyapunov function for (4.5), we need to take into account the specific forms of the utility functions U_s. Take for example the class of (p, α)-fair utility functions [1]:

$$U_s(x_s) = \begin{cases} p_s \frac{(x_s)^{1-\alpha}}{1-\alpha}, & 0 \le \alpha < 1, \\ p_s \log(x_s), & \alpha = 1, \\ p_s \log\left(\frac{x_s}{x_s+1}\right), & \alpha = 2, \\ p_s[\log\left(\frac{x_s}{x_s+1}\right) + \sum_{i=1}^{\alpha-2} \frac{1}{i(x_s+1)^i}], & \alpha = 3, 4, \ldots \end{cases} \qquad (4.6)$$

This class of utility functions were widely adopted in the NUM paradigm [2, 3]. The proportional-fair utility function in Chap. 2 is in fact the $(p, 1)$-fair function. Furthermore, maximizing the utility function (4.6) corresponds to maximizing weighted throughput when $\alpha = 0$, and achieving max-min fairness as $\alpha \to \infty$. It was shown that use of such $U_s(x_s)$ as the objective functions can nicely trade off the throughput and fairness; adoption of larger α leads to more fairness.

For $U_s(x_s)$ in (4.6) with $0 \le \alpha \le 1$, we can establish that:

Theorem 4.1 *Under the scheduler (2.11), the window-based control (4.4) with $1 - \alpha \le \rho \le 1$ globally converges to its unique equilibrium that leads to the optimal x^* for (2.1) with $U_s(x_s)$, $0 \le \alpha \le 1$, in (4.6).*

Proof See Appendix 4D.

For $U_s(x_s)$ in (4.6) with $\alpha \ge 1$, we can also show that:

Theorem 4.2 *Under the scheduler (2.11), the window-based control (4.4) with $0 \le \rho \le 1$ globally converges to its unique equilibrium that leads to the optimal x^* for (2.1) with $U_s(x_s)$, $\alpha \ge 1$, in (4.6), if we select $p_s < \frac{d_s x_s^{-\alpha}}{\alpha - 1}$, $\forall s$.*

Proof See Appendix 4E.

Both Theorems 4.1 and 4.2 are generalizations of Theorem 2.2. For (p, α)-fair utility functions $U_s(x_s)$ with $0 \leq \alpha \leq 1$, we need use a $\rho \geq 1 - \alpha$ ($\rho \geq 0$ for $\alpha = 1$) in (4.5) to warrant its global convergence to the optimal equilibrium. On the other hand, for $U_s(x_s)$ with $\alpha \geq 1$, we need select sufficiently small $p_s < \frac{d_s x_s^{-\alpha}}{\alpha - 1}$ ($p_s < \infty$ for $\alpha = 1$), $\forall s$, to ensure the global convergence of (4.5).

Based on (4.5), QUIC-TCP algorithm (2.13) can be modified such that it together with the link-scheduler (2.14) can entail a window-based implicit primal-dual solver for the NUM problem (2.1) with any given (p, α)-fair utility $U_s(x_s)$. With careful designs, these schemes are readily deployable and can have performance guarantees per Theorems 4.1 and 4.2. Related insights can be employed to re-develop numerous existing NUM cross-layer optimization algorithms into high-performance practical schemes that can be deployed and operated over current Internet infrastructure.

4.3 Interesting Directions

The proposed approach clearly has many attractive features with far reaching implications and a host of interesting directions to build on. We next outline a couple of them.

4.3.1 Impact of Feedback Delays and Network Dynamics

Development and analysis of the proposed schemes in Chaps. 2 and 3 assumed a fluid model where the feedback delays and stochastic traffic/channel dynamics are absent. Although convergence and stability of the proposed network schemes in the presence of feedback delays were demonstrated by our ns-2 simulations [4, 5], it is still important to explore such a convergence/stability analysis to provide performance guarantees in realistic network environment.

Consider the impact of feedback delays. Given the round-trip delay $\bar{d}_s(t) = d_s + q^s(t)$, the equilibrium of the window adjustment (2.12) should maintain: $v_s(t) = w_s(t) - x_s(t - \bar{d}_s(t))d_s - p_s = 0$, $\forall s$, when feedback delays are taken into account. Accordingly, the window control would become:

$$\frac{d}{dt} w_s(t) = -\frac{d_s}{d_s + q^s(t)} w_s(t)^{-2\rho + 1} \left(w_s(t) - \frac{w_s(t - d_s - q^s(t))}{d_s + q^s(t)} d_s - p_s \right).$$

Denote the forward feedback delay from source s to link l as τ_{ls}^f and the backward feedback delay from link l to source s as τ_{ls}^b. Then the aggregate queueing delays $q^s(t) = \sum_{l \in L(s)} q_l(t - \tau_{ls}^b)$. On the other hand, the complimentary

conditions between queueing delays and link capacity constraints become: $q_l(t)$
$\left(c_l - \sum_{s \in S(l)} \frac{w_s(t-\tau_{ls}^f)}{d_s+q^s(t-\tau_{ls}^f)}\right) = 0$ or $q_l(t)\left(r_l(t) - \sum_{s \in S(l)} \frac{w_s(t-\tau_{ls}^f)}{d_s+q^s(t-\tau_{ls}^f)}\right) = 0$; and e.g.,
the queueing-delay based MaxWeight scheduler is to find: $r(t) := \arg\max_{r \in \bar{\mathscr{R}}}$
$\sum_{l \in L_w} q_l(t)r_l$.

Stability analysis of the foregoing feedback control system is challenging. A viable direction is to study its local stability based on the generalized Nyquist criterion [6, 7]; see such analysis in [8–11] for TCP over wired Internet. To show local stability of the proposed schemes, we can linearize the system around its equilibrium. Taking the Laplace transform for the linearized system with feedback delays, we could derive the open loop transfer function $\mathscr{L}(s)$ for the proposed window adjustment mechanism. Implied by the Nyquist criterion, this linearized system is stable, and thus the original system is locally stable, if the spectral radius of $\mathscr{L}(e^{j\omega})$ is strictly less than 1 for $\omega \in (0, 2\pi)$. Using this criterion, we can derive the sufficient conditions for locally asymptotic stability of the proposed network schemes in the presence of feedback delays. The stability analysis can reveal the impact of feedback delays on performance of the proposed cross-layer optimization.

The deterministic fluid model analysis also ignores the system dynamics. Performance of the proposed schemes under stochastic channel and network traffic dynamics needs to be carefully investigated. Although the robustness of the proposed schemes to network dynamics was shown by simulations, analytical convergence/optimality claims are called for. For the queue-length based NUM schemes, near-optimality and convergence under channel and traffic dynamics were established via stochastic approximation tools, e.g., Stolyar's fluid limit argument [12], Eryilmaz's Foster-Lyapunov method [13], Neely's Lyapunov optimization framework [14], and our stochastic averaging approach [15].

We can adopt these techniques to analyze the proposed queueing-delay based schemes in dynamic network models. Building on the fluid model analysis, Stolyar's fluid limit argument could be applied to account for channel and traffic dynamics with the proposed algorithms, and then convergence of the proposed schemes (in probability) under network dynamics readily follow. How to take appropriate time scaling to get the fluid limit and what it means for the real model are in fact quite subtle. Other stochastic analysis approaches such as those in [13–15] and the "stochastic approximation with controlled Markov noise" argument in [16] can be also employed. Relying on these tools, a unifying stochastic analysis framework for the proposed network schemes can be developed for wireless Internet applications in general stochastic channel/traffic models. The analysis can provide guidelines for implementation of the proposed schemes under network dynamics.

4.3.2 Multi-Path Routing

In formulation (2.1), each flow is assigned a single path between its source and destination. This is consistent with the current Internet routing protocol which selects only a single route per flow. However, single-path routing may lead to congestion due to inefficient routing decision itself, and thus limit the TCP throughput, especially for the wireless networks [17, 18]. It is well motivated to consider networks where multiple paths are available to each flow and the source can direct its flow along these paths using source routing [19]. Currently, source routing is not supported in the Internet. Yet, it is envisioned that this scheme could be allowed to exploit path diversity in future overlay and multi-homing Internet applications; see Fig. 4.2.

We can generalize the proposed approach to develop readily deployable and optimal network control algorithms for networks supporting multipath source routing. This requires a joint congestion control and source-routing design, which was considered for packet-loss based TCP in wired networks [9]. Building on the related insights, we can develop joint congestion control, routing, and link-scheduling schemes for wireless Internet applications within our design-space oriented cross-layer optimization framework. To this end, we need to reformulate the optimization problem in (2.1). With source routing, each flow s can be split among available routes $R(s)$ subject to the link capacity constraints, and the desirable congestion control algorithms should be able to automatically find such a split while ensuring congestion avoidance. For each flow s, let x_m denote the source rate on a route $m \in R(s)$; hence, the total source rate is $\sum_{m \in R(s)} x_m$. Consider to maximize the aggregate utility:

$$U(x) := \sum_{s \in S} \left(p_s \log \left(\sum_{m \in R(s)} x_m \right) + \varepsilon \sum_{m \in R(s)} \log x_m \right).$$

Here different from the weighted proportionally-fair utility function in (2.1), we add a term containing a small ε to ensure that the utility function is strictly concave; hence, it has a unique maximizer [9]. Clearly, such a utility approaches the proportionally-fair one for ε sufficiently small. As each flow has multiple routes, the link capacity constraints should be enforced with respect to routes. Let $M(l)$ denote set of the routes that use link l. The link capacity constraints become:

Fig. 4.2 Multi-homed server: One TCP connection with two paths

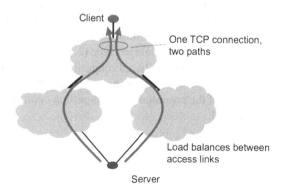

$$\sum_{m \in M(l)} x_m \leq c_l, \quad \forall l \in L_f,$$

$$\sum_{m \in M(l)} x_m \leq r_l, \quad \forall l \in L_w.$$

Let $L(m)$ denote the set of links that route m uses, and $q^m = \sum_{l \in L(m)} q_l$ the aggregate queueing delay per route m. With queueing delays playing the role of Lagrange multipliers, the KKT condition of the reformulated problem then implies that

$$p_s \Big/ \sum_{m' \in R(s)} x_{m'} + \varepsilon / x_m = q^m, \quad \forall m \in R(s).$$

Let w_m be the window size and d_m the propagation delay for route m. Under the fluid model, it also holds that $x_m = w_m/(d_m + q^m)$. Therefore, the optimal window sizes should render:

$$v_m = w_m - x_m d_m - \varepsilon - p_s x_m \Big/ \sum_{m' \in R(s)} x_{m'} = 0, \quad \forall m \in R(s),$$

at each source s. As the formulated problem has a unique optimal x^*, it can be shown that the optimal window vector w^* is unique. Since the values of $v_m, \forall m \in R(s)$, can be obtained for each source s by its own observation/estimation, $v_m = 0$ can serve as decoupled optimality criteria for the sources.

As with (2.12), we can then develop end-to-end window adjustment according to an ODE:

$$\frac{d}{dt} w_m(t) = -g_m(w_s, d_s, x_s) v_m, \quad \forall m \in R(s), \forall s,$$

where $g_m(w_s, d_s, x_s)$ is a function of *local* vectors $w_s := \{w_m, \forall m \in R(s)\}$, $d_s := \{d_m, \forall m \in R(s)\}$, and $x_s := \{x_m, \forall m \in R(s)\}$. Lyapunov analysis can be carried out to address the convergence of the proposed window control. The suitable Lyapunov functions can be identified, and the specific form of the function $g_m(w_s, d_s, x_s)$ can be in turn determined. Under the queueing-delay based link-scheduler e.g., the MaxWeight-type (2.11), the global or local convergence/stability of the proposed window-based congestion control and source-routing algorithm to its unique equilibrium can be then established. Based on the analysis, practical implementation of the proposed joint congestion control and source-routing schemes can be proposed.

4.4 Summary

We outline some possible generalizations and interesting directions for the novel design-space oriented cross-layer optimization approach developed in Chaps. 2 and 3. We generalize our approach to the Internet with both cellular and ad-hoc wireless links and to the NUM with general concave utility functions. In addition, we outline that generalized Nyquist criterion and stochastic approximation tools can be employed to study the stability and performance of the proposed network schemes in realistic Internet environments with feedback delays and network traffic/channel dynamics, and the proposed approach can be also generalized to joint congestion control, routing, and link-scheduling optimization for networks supporting multi-path source routing.

Appendix 4A: Proof of Corollary 4.1

The KKT conditions imply that optimal $\{x^*, r^*, u^*\}$ for (4.1) and optimal λ^* for its dual problem satisfy [20]:

$$\frac{p_s}{x_s^*} = \lambda^{s*} := \sum_{l \in L(s)} \lambda_l^*, \quad \forall s \in S \tag{4.7}$$

$$\lambda_l^* \left(c_l - \sum_{s \in S(l)} x_s^* \right) = 0, \quad \sum_{s \in S(l)} x_s^* \leq c_l, \quad \forall l \in L_f \tag{4.8}$$

$$\lambda_l^* \left(r_l^* - \sum_{s \in S(l)} x_s^* \right) = 0, \quad \sum_{s \in S(l)} x_s^* \leq r_l^*, \quad \forall l \in L_c \tag{4.9}$$

$$\lambda_l^* \left(\sum_{s \in S(l)} x_s^* - b_l \sum_i [u_i^* \xi_l^i] \right) = 0, \quad \sum_{s \in S(l)} x_s^* \leq b_l \sum_i [u_i^* \xi_l^i], \quad \forall l \in L_a \tag{4.10}$$

$$r^* = \arg\max_{r \in \mathcal{R}} \sum_{l \in L_c} \lambda_l^* r_l, \quad u_i^* = \frac{\exp(\sum_{l \in L_a} (\lambda_l^* b_l \xi_l^i))}{\sum_j \exp(\sum_{l \in L_a} (\lambda_l^* b_l \xi_l^j))}, \quad \forall i. \tag{4.11}$$

On the other hand, we have the following relationships for source rates x_s, window sizes w_s, and queueing delays q_l in the network fluid model:

$$x_s (d_s + q^s) = w_s, \quad \forall s \tag{4.12}$$

$$q_l \left(C_l - \sum_{s \in S(l)} x_s \right) = 0, \quad \sum_{s \in S(l)} x_s \leq C_l, \quad \forall l \tag{4.13}$$

$$\text{where } C_l = \begin{cases} c_l & \forall l \in L_f \\ r_l & \forall l \in L_c \\ R_l & \forall l \in L_a \end{cases}$$

At the equilibrium $v = 0$ we have $w_s^* - x_s^* d_s - p_s = 0$, and it follows from (4.12) that $w_s^* = x_s^*(d_s + q^{s*})$. Hence, we readily have $p_s = x_s^* q^{s*}$, $\forall s$. This is exactly the KKT condition (4.7) if we let $\lambda^* \equiv q^*$. With this equivalence mapping, moreover, the fluid model identities (4.13), (2.11), and (3.13) for $\{x^*, r^*, u^*, q^*\}$ also become the KKT conditions (4.8–4.11). This then implies that x^*, r^* and q^* are the optimal solutions of (4.1) and its dual problem. Furthermore, we can show that the optimal window equilibrium w^* always exists and it is unique due to the existence and uniqueness of x^* for (4.1), as well as the uniqueness of mapping $w^* \to x^*$.

Appendix 4B: Proof of Corollary 4.2

For a given w, let $\mathscr{B}$ denote the set of bottleneck links. Define $A_B, q_B, c_B,$ and R_B the corresponding sub-matrix or sub-vectors of q, c, and R for bottleneck links. Let the diagonal matrices $X := \text{diag}(x)$, $W := \text{diag}(w)$, $D := \text{diag}(d)$, and $\bar{D} := \text{diag}(\bar{d})$ where $\bar{d} := \{\bar{d}_s, \forall s\}$. We have:

$$X(A_B^T q_B + d) = w. \tag{4.14}$$

Differentiating both sides of (4.14) with respect to w and then multiplying both sides by $A_B \bar{D}^{-1}$ yield:

$$A_B J_{x|w} + A_B \bar{D}^{-1} X A_B^T J_{q_B|w} = A_B \bar{D}^{-1}. \tag{4.15}$$

Without loss of generality, consider now a bottleneck set containing all wireless links. Partition the bottleneck-only routing matrix A_B and queueing delay vector q_B into three parts:

$$A_B = \begin{bmatrix} A_{f,B} \\ A_c \\ A_a \end{bmatrix}, \qquad q_B = \begin{bmatrix} q_{f,B} \\ q_c \\ q_a \end{bmatrix}$$

where subscripts f,B, c, and a denote the parts related to the wired, centralized, and ad-hoc wireless bottleneck links, respectively.

For the wired bottleneck links, it holds $A_{f,B} x = c_B$, and thus $A_{f,B} J_{x|w} = 0$. On the other hand, it holds for wireless bottleneck links: $A_c x = r^*$ and $A_a x = R$. This implies:

$$A_c J_{x|w} = [0 \quad J_{r^*|q_c} \quad 0] J_{q_B|w},$$
$$A_a J_{x|w} = [0 \quad 0 \quad J_{R|q_a}] J_{q_B|w}.$$

Then overall we have

$$A_B J_{x|w} = \begin{bmatrix} A_{f,B} \\ A_c \\ A_a \end{bmatrix} \quad J_{x|w} = \begin{bmatrix} 0 & 0 & 0 \\ 0 & J_{r^*|q_c} & 0 \\ 0 & 0 & J_{R|q_a} \end{bmatrix} \quad J_{q_B|w} := N J_{q_B|w}.$$

It then follows from (2.18) that

$$(N + A_B \bar{D}^{-1} X A_B^T) J_{q_B|w} = A_B \bar{D}^{-1}.$$

Since both $J_{r^*|q_c}$ and $J_{R|q_a}$ are positive definite from Lemmas 2.2 and 3.1, the matrix N is clearly positive semi-definite, and thus $N + A_B \bar{D}^{-1} X A_B^T$ is positive definite. Using the convenient notation: $M := A_B^T(N + A_B \bar{D}^{-1} X A_B^T)^{-1} A_B$, it is clear that the matrix M is positive semi-definite and $J_{x|w} = \bar{D}^{-1}(I - X M \bar{D}^{-1})$.

With $Y(w) = \frac{1}{2} \sum_{s \in S} \left(\frac{v_s}{(w_s)^\rho} \right)^2$, we can then follow the similar lines in the proof of Theorem 2.2 to show $dY(w(t))/dt < 0$, i.e., $Y(w(t))$ is strictly decreasing in t, at all points unless $v = 0$. The unique equilibrium $v = 0$ is thus globally asymptotically stable.

Appendix 4C: Proof of Proposition 4.1

Define the diagonal matrices $U' := \text{diag}(\{U_s'(x_s), \ \forall s\})$ and $U'' := \text{diag}(\{U_s''(x_s), \ \forall s\})$. Then $v = w - Dx - U'x$; consequently,

$$J_{v|w} = I - (D + U' + U''X) J_{x|w} = I - (D + U' + U''X) \bar{D}^{-1}(I - X M \bar{D}^{-1}).$$

Let $\tilde{U} := D + U' + U''X$ and define the diagonal matrix $G := \text{diag}(\{g_s(w_s, d_s, x_s), \ \forall s\})$. With $Y(w) = \frac{1}{2} \sum_{s \in S} \left(\frac{v_s}{(w_s)^\rho} \right)^2$, we have [cf. (2.20)]:

$$\begin{aligned}
\frac{d}{dt} Y(w(t)) &= \sum_s \left(\frac{\partial Y}{\partial w_s} \frac{dw_s(t)}{dt} \right) \\
&= -v^T W^{-\rho} [W^{-\rho} J_{v|w} - \rho W^{-\rho-1} V] G v \\
&= -v^T [W^{-2\rho}(I - \tilde{U} \bar{D}^{-1}(I - X M \bar{D}^{-1})) - \rho W^{-2\rho-1}(W - DX - U'X)] G v \\
&= -v^T [(1 - \rho) W^{-2\rho}(I - D\bar{D}^{-1}) + \rho W^{-2\rho-1} U'X - W^{-2\rho}(U' + U''X) \bar{D}^{-1} \\
&\quad + W^{-2\rho} \tilde{U} \bar{D}^{-1} X M \bar{D}^{-1}] G v. \tag{4.16}
\end{aligned}$$

To render $\frac{d}{dt}Y(w(t)) < 0, \forall v \neq 0$, we require the last term $v^T W^{-2\rho}\tilde{U}\bar{D}^{-1}XM\bar{D}^{-1}Gv \geq 0$. For the semi-definite matrix M, it is necessary that $G = X\tilde{U}W^{-2\rho}$; i.e.,

$$g_s(w_s, d_s, x_s) = x_s(d_s + U_s'(x_s) + x_s U_s''(x_s))w_s^{-2\rho}, \quad \forall s.$$

Appendix 4D: Proof of Theorem 4.1

For $U_s(x_s)$ in (4.6) with $0 \leq \alpha \leq 1$, we have $U_s'(x_s) = p_s x_s^{-\alpha}$, $U_s''(x_s) = -\alpha p_s x_s^{-\alpha-1}$, and $g_s(w_s, d_s, x_s) = x_s[d_s + (1-\alpha)\frac{p_s}{x_s^\alpha}]w_s^{-2\rho}$ in (4.5). Substituting these terms into (4.16), we have:

$$\begin{aligned}
\frac{d}{dt}Y(w(t)) = -v^T\{&[(1-\rho)W^{-2\rho}(I - D\bar{D}^{-1}) \\
&+ (\rho + \alpha - 1)W^{-2\rho}PX^{-\alpha}\bar{D}^{-1}]X\tilde{U}W^{-2\rho} \\
&+ W^{-2\rho}\tilde{U}X\bar{D}^{-1}M\bar{D}^{-1}X\tilde{U}W^{-2\rho}\}v
\end{aligned}$$

where $\tilde{U} = D + (1-\alpha)PX^{-\alpha}$. Since $(I - D\bar{D}^{-1})$ is a diagonal matrix with nonnegative entries, M is positive semi-definite, and all W, D, $\bar{D}$, P, and $\tilde{U}$ are diagonal matrices with positive diagonal entries, the whole matrix inside the curvy bracket is positive definite for $1-\alpha \leq \rho \leq 1$. Hence, $dY(w(t))/dt < 0$, i.e., $Y(w(t))$ is strictly decreasing in t unless $v = 0$.

Appendix 4E: Proof of Theorem 4.2

For $U_s(x_s)$ in (4.6) with $\alpha \geq 1$, we have $U_s'(x_s) = \frac{p_s}{x_s(1+x_s)^\alpha}$, $U_s''(x_s) = -p_s\frac{1+(1+\alpha)x_s}{x_s^2(1+x_s)^{\alpha+1}}$, and $g_s(w_s, d_s, x_s) = x_s[d_s - \frac{\alpha p_s}{(1+x_s)^{\alpha+1}}]w_s^{-2\rho}$ in (4.5); hence,

$$\tilde{U} = D + U' + U''X = D - \alpha P(I + X)^{-\alpha-1}.$$

All the diagonal entries of $\tilde{U}$ are positive if we select $p_s < \frac{d_s x_s^{-\alpha}}{\alpha-1}$, $\forall s$. In this case, substituting $G = X\tilde{U}W^{-2\rho}$ into (4.16) leads to:

$$\begin{aligned}
\frac{d}{dt}Y(w(t)) = -v^T\{&[(1-\rho)W^{-2\rho}(I - D\bar{D}^{-1}) + \alpha W^{-2\rho}(I + X)^{-\alpha-1}\bar{D}^{-1} \\
&+ \rho W^{-2\rho-1}(I + X)^{-\alpha}]X\tilde{U}W^{-2\rho} \\
&+ W^{-2\rho}\tilde{U}X\bar{D}^{-1}M\bar{D}^{-1}X\tilde{U}W^{-2\rho}\}v
\end{aligned}$$

Again, the whole matrix inside the curvy bracket is positive definite for $0 \le \rho \le 1$, if we select $p_s < \frac{d_s x_s^{-\alpha}}{\alpha - 1}$, $\forall s$. Hence, $dY(w(t))/dt < 0$, i.e., $Y(w(t))$ is strictly decreasing in t unless $v = 0$.

References

1. J. Mo, J. Walrand, Fair end-to-end window-based congestion control. IEEE/ACM Trans. Netw. **8**(5), 556–567 (2000)
2. M. Chiang, S. Low, A. Calderbank, J. Doyle, Layering as optimization decomposition. Proc. IEEE **95**(1), 255–312 (2007)
3. D. Palomar, M. Chiang, A tutorial on decomposition methods for network utility maximization. IEEE J. Sel. Areas Commun. **24**(8), 1439–1451 (2006)
4. X. Wang, Z. Li, N. Gao, Joint congestion control and wireless-link scheduling for mobile TCP applications, *Proc. of IEEE Globecom Conf.*, Houston, TX, Dec. 5–9, 2011.
5. X. Wang, Z. Li, Joint optimization of TCP congestion control and distributed CSMA scheduling, *Proceedings of IEEE Globecom Conference*, Anaheim, CA, 3–7 Dec. 2012
6. C. Desoer, Y. Wang, On the generalized Nyquist stability criterion. IEEE Trans. Autom. Control **25**(2), 187–196 (1980)
7. F. Callier, C. Desoer, *Linear System Theory* (Springer, New York, 1991)
8. F. Kelly, A. Maulloo, D. Tan, Rate control in communication networks: shadow prices, proportional fairness and stability. J. Oper. Res. Soc. **49**(3), 237–252 (1998)
9. H. Han, S. Shakkottai, C. Hollot, R. Srikant, D. Towsley, Multi-path TCP: a joint congestion control and routing scheme to exploit path diversity in the Internet. IEEE/ACM Trans. Netw. **14**(6), 1260–1271 (2006)
10. J. Wang, A. Tang, S. Low, Local stability of FAST TCP, *Proceedings of IEEE Conference on Decision and Control*, pp. 1023–1028, Dec. 2004
11. J. Wang, D. Wei, S. Low, Modeling and stability of FAST TCP, *Proceedings of IEEE INFOCOM Conference*, pp. 938–948, Miami, FL, 13–17 Mar. 2005
12. A. Stolyar, Maximizing queueing network utility subject to stability: greedy primal-dual algorithm. Queueing Syst. **50**, 401–457 (2005)
13. A. Eryilmaz, R. Srikant, Fair resource allocation in wireless networks using queue-length-based scheduling and congestion control, *Proceedings of IEEE INFOCOM Conference*, vol. 3, pp. 1794–1803, Miami, FL, 13–17 Mar. 2005
14. M. Neely, E. Modiano, C. Li, Fairness and optimal stochastic control for heterogenous networks. IEEE/ACM Trans. Netw. **16**(2), 396–409 (2008)
15. X. Wang, G.B. Giannakis, A.G. Marques, A unified approach to QoS-guaranteed scheduling for channel-adaptive wireless networks. Proc. IEEE **95**(12), 2410–2431 (2007)
16. J. Liu, Y. Yi, A. Proutiere, M. Chiang, H. Poor, Towards utility-optimal random access without message passing, *Wireless Communications and Mobile Computing*. doi:10.1002/wcm.000
17. S. Mueller, R. Tsang, D. Ghosal, Multipath routing in mobile ad hoc networks: issues and challenges, in *Performance Tools and Applications to Networked Systems*, vol. 2965, ed. by M. Calzarossa, E. Gelenbe (Springer, Berlin, 2004), pp. 209–234
18. S. Bahk, M. El Zarki, Dynamic multi-path routing and how it compares with other dynamic routing algorithms for high speed wide area network. ACM SIGCOMM Comput. Commun. Rev. **22**(4), 53–64 (1992)
19. D. Johnson, D. Maltz, Dynamic source routing in ad hoc wireless networks. Mobile Comput. (The International Series in Engineering and Computer Science) **353**, 153–181 (1996)
20. S. Boyd, L. Vandenberghe, *Convex Optimization* (Cambridge University Press, Cambridge, 2004)

Chapter 5
Summary

We develop a novel *design-space oriented* cross-layer optimization paradigm for wireless Internet applications. Different from the existing theory-guided design methods, the main ideas behind our approach are the use of queueing delays to regulate the protocol interactions across layers, and development of window-based cross-layer optimization algorithms. Confined to the Internet design space, the proposed cross-layer optimization approach is then to develop non-standard *window-based implicit primal-dual solvers* for underlying optimization problems, and rely on queueing delays to *decompose* these solvers into local algorithms that can be deployed and operated asynchronously at different layers of distributed network nodes. With a focus on joint congestion control and link-scheduling design for wireless Internet applications, the book establishes the feasibility of this approach by showing the existence of such solvers and creating a systematic framework for their design and analysis.

The proposed novel cross-layer optimization approach is an attempt to bridge the network optimization theory and practical Internet designs, by developing non-standard optimization algorithms within the Internet design space towards network protocols/schemes that can be used with current infrastructure. On the scientific front, the approach is possible to embody a major paradigm shift in the theory and practice for cross-layer design and optimization of Internet protocols. As a result, the research can impact the *basic theory* in understanding and optimizing the large-scale network and communication systems, and is expected to benefit directly *applications* to next-generation Internet protocol designs.

On the social front, the proposed readily deployable, scalable yet jointly optimal network schemes can greatly improve the performance of the emerging wireless Internet applications. In a broad sense, these high-performance schemes can have an *impact on society*, if one takes into account how the Internet and wireless devices (such as Apple iPhones, iPads and Android Phones) have transcended various aspects of our everyday life: from work environments to "home network", where "smart devices" enable wireless Internet access among phones, equipments and appliances to assist in performing tasks with limited human intervention.

X. Wang, *Scheduling and Congestion Control for Wireless Internet*,
SpringerBriefs in Electrical and Computer Engineering,
DOI: 10.1007/978-1-4614-8420-2_5, © The Author(s) 2014